Essentials of Pharmaceutical Preformulation

Essentials of Pharmaceutical Preformulation

Simon Gaisford
UCL School of Pharmacy, University College London, London, UK

Mark Saunders
Kuecept Ltd, Potters Bar, Herts, UK

WILEY-BLACKWELL

A John Wiley & Sons, Ltd., Publication

This edition first published 2013
© 2013 John Wiley & Sons, Ltd

Registered office
John Wiley & Sons Ltd, The Atrium, Southern Gate, Chichester, West Sussex, PO19 8SQ,
United Kingdom

For details of our global editorial offices, for customer services and for information about how to
apply for permission to reuse the copyright material in this book please see our website at
www.wiley.com.

Library of Congress Cataloging-in-Publication Data

Gaisford, Simon.
  Essentials of pharmaceutical preformulation / Simon Gaisford and Mark Saunders.
    p. ; cm.
  Includes bibliographical references and index.
  ISBN 978-0-470-97635-7 (cloth) – ISBN 978-0-470-97636-4 (paper)
  I. Saunders, Mark, 1976 Sept. 18-   II. Title.
  [DNLM: 1. Drug Compounding–methods.   2. Drug Discovery–methods. QV 779]
  615.1′9–dc23
                                                              2012027538

A catalogue record for this book is available from the British Library.

Wiley also publishes its books in a variety of electronic formats. Some content that appears in print
may not be available in electronic books.

Cover design: Gary Thompson

Set in 10.5/13pt Times Ten by Aptara Inc., New Delhi, India

First Impression   2013

For Yasmina and Oliver

Contents

Companion website

This book is accompanied by a companion website at:

http://www.wiley.com/go/gaisford/essentials

Visit the website for:
- Figures and tables from the book
- Multiple choice questions

Preface

The inspiration for this book came when Michael Aulton asked me to take over his preformulation module on the PIAT course of the University of Manchester. The existing module was based on the excellent textbook (*Pharmaceutical Preformulation*) written by Jim Wells in 1988 and thus a perfect opportunity to write both an updated module and an updated book presented itself.

The majority of the text was written while I was on sabbatical leave at the Monash Institute for Pharmaceutical Sciences (MIPS) in Melbourne, Australia, in the summer of 2011. I am hugely grateful to Prof. Bill Charman, Prof. Peter Stewart, Marian Costelloe and Marian Glennon for arranging the visit and to MIPS as a whole for the welcoming, friendly and stimulating environment they provided. I would also like to mention many of the wonderful people that I met, including Anne, Ben, Carolyn, Chris, Colin, David, Ian, Hywel, Iliana, Joe, Laurence, Mercedes and Michelle. Special thanks are reserved for Richard Prankerd, who took the time and effort to talk with me about many aspects of the text while I was at MIPS and also to review the whole text before publication; the book is immeasurably better for his input and advice. Thermal analysts are indeed a special breed!

Equally, the book would never have been finished were it not for several other special people. Nicole Hunter undertook the weighty tasks of reviewing the whole text and providing constant support and encouragement while Hamid Merchant cast his expert eye over the dissolution chapter. My group of amazing PhD students (Alice, Asma, Garima, Jawal, Jip, Luis, Mansa, Mustafa and Rin) provided many of the data and examples that populate the text while my academic colleagues, particularly Prof. Abdul Basit, Prof. Anthony Beezer and Prof. Kevin Taylor, have been constant sources of advice and support. I also acknowledge all of the wonderful students who I have taught on the MSc in Drug Delivery and who are such an inspiration to me.

Of course, no book would be possible without a publisher, and I am extremely grateful to Fiona Seymour and Lucy Sayer for editorial advice and encouragement.

Finally, I must acknowledge the constant support of my family, especially Joanne and Oliver, who keep me sane!

S Gaisford
April 2012

List of Abbreviations

Abbreviation	Meaning
ε	Molar absorption coefficient
θ	Angle of repose
σ	Normal stress
τ	Shear stress
A	Surface area
AFM	Atomic force microscopy
AR	Aspect ratio
BCS	Biopharmaceutical Classification System
BP	British Pharmacopoeia
C	Concentration
CRM	Certified reference material
D	Diffusion coefficient
DSC	Differential scanning calorimetry
ESEM	Environmental scanning electron microscopy
F	Dilution factor
FaSSIF	Fasted state simulated intestinal fluid
FDA	US Food and Drug Administration
FeSSGF	Fed state simulated gastric fluid
FeSSIF	Fed state simulated intestinal fluid
FTIR	Fourier transform infrared
G	Gibb's free energy
GRAS	Generally regarded as safe
h	Thickness of boundary layer
H	Enthalpy
HPLC	High-performance liquid chromatography
HSM	Hot-stage microscopy
ICH	International Conference on Harmonisation

IDR	Intrinsic dissolution rate
IR	Infrared
IUPAC	International Union of Pure and Applied Chemistry
k	Rate constant
K	Equilibrium constant or stress ratio
MTDSC	Modulated temperature DSC
n	Reaction order
NA	Numerical aperture
NIR	Near infrared
p	Negative logarithm
PhEur	European Pharmacopoeia
RI	Refractive index
S	Entropy
S_0	Intrinsic solubility
SEM	Scanning electron microscopy
SGF	Simulated gastric fluid
T	Temperature
TEM	Transmission electron microscopy
TGA	Thermogravimetric analysis
TLC	Thin-layer chromatography
UHPLC	Ultra high-performance liquid chromatography
USP	United States Pharmacopoeia
UV	Ultraviolet
V	Volume
W	Weight
x	Mole fraction
XRPD	X-ray powder diffraction

1
Basic Principles of Preformulation Studies

1.1 Introduction

The worldwide market for pharmaceutical sales is large and has grown consistently year-on-year for much of the past decade (Table 1.1). The advent of computer-based drug design programmes, combinatorial chemistry techniques and compound libraries populated with molecules synthesised over many decades of research and development means there is a vast array of compounds with the potential to become drug substances. However, drug substances are not administered to patients as pure compounds; they are formulated into drug products. The selection of a compound, its development into a drug substance and, ultimately, drug product is a hugely time-consuming and expensive process, which is ultimately destined for failure in the majority of cases. As a rough guide, only 1 out of every 5–10 000 promising compounds will be successfully developed into a marketed drug product and the costs involved have been estimated at ca. $1.8 billion (Paul *et al.*, 2010).

While it is tempting to assume that all drug products are financial blockbusters, approximately 70% never generate sufficient sales to recoup their development costs. Table 1.2 shows the top 20 medicines by sales worldwide (and the percentage of revenue they generate for their respective companies). It is apparent that a significant percentage of income is generated from these blockbuster products, and the financial health and prospects of the originator company are largely dependent upon the extent of patent protection (allowing market exclusivity) and new drug products in the development pipeline.

These numbers imply that development of a drug product in the right therapeutic area can result in significant income, but the costs involved in

---

Essentials of Pharmaceutical Preformulation, First Edition. Simon Gaisford and Mark Saunders.
© 2013 John Wiley & Sons, Ltd. Published 2013 by John Wiley & Sons, Ltd.

Table 1.1   Total market sales in the pharmaceutical sector from 2003 to 2010 (data from IMS Health).

	2003	2004	2005	2006	2007	2008	2009	2010
Total market[a]	500	560	605	651	720	788	819	856
% Growth	9.1	7.6	7.2	7.0	6.9	6.1	7.1	4.1

[a]US$ in billions.

reaching market are such that only a few potential drug substances can be considered for development. How best to select a compound for development from the myriad of chemical structures that may be available? It is tempting to think that the decision reduces to efficacy against a biological target alone, but in practice physicochemical properties affect how a substance will process, its stability and interaction with excipients, how it will transfer to solution and, ultimately, define its bioavailability. The compound showing greatest efficacy may not ultimately be selected if another compound has a better set of physicochemical properties that make it easier to formulate and/or manufacture. It follows that characterising the physicochemical properties of drug substances early in the development process will provide the fundamental knowledge base upon which candidate selection, and in the limit dosage form design, can be made, reducing development time and cost. This is the concept of preformulation.

1.2   Assay design

In the early stages of preformulation the need rapidly to determine bioavailability, dose and toxicity data predominate and hence the first formulations

Table 1.2   Top ten drugs by sales worldwide in 2010 (data from IMS Health).

Product	Manufacturer	Sales (US$m)	% of company sales	Date of patent expiry
Lipitor	Pfizer	12 657	22.8	2011
Plavix	Sanofi-Aventis/BMS	8817	17.3[a]	2012
Seretide	GlaxoSmithKline	8469	25.2	2013[b]
Nexium	AstraZeneca	8362	23.5	2014
Seroquel	AstraZeneca	6816	19.2	2012
Crestor	AstraZeneca	6797	19.1	2012
Enbrel	Amgen/Pfizer	6167	8.7[a]	2012
Remicade	Janssen/Schering-Plough	6039	n/d	2011
Humira	Abbott	5960	25.0	2016
Zyprexa	Eli Lilly	5737	25.9	2011

[a]Based on combined sales of both companies.
[b]European expiry. The US patent expired for Seretide in 2010.

Table 1.3 Molecular sample properties and the assays used to determine them.

Property	Assay	Requirement of sample
Solubility[a] • Aqueous • Nonaqueous	UV	Chromophore
pK_a	UV or potentiometric titration	Acid or basic group
$P_{o, w}$/log P	UV TLC HPLC	Chromophore
Hygroscopicity	DVS TGA	No particular requirement
Stability • Hydrolysis • Photolysis • Oxidation	HPLC, plus suitable storage conditions	No particular requirement

[a]Solubility will depend on physical form.

of a drug substance are usually for intravenous injection. The first task facing any formulator is thus to prepare a suitable formulation for injection – most often this requires only knowledge of solubility and the development of a suitable assay. It is extremely important to note here that no development work can proceed until there is a suitable assay in place for the drug substance. This is because experimentation requires measurement.

1.2.1 Assay development

Assays greatly assist quantitative determination of physicochemical parameters. Since each assay will in general be unique to each drug substance (or, more correctly, *analyte*) development of assays may be time-consuming in cases where many drug substances are being screened. The first assays developed should ideally require minimum amounts of sample, allow determination of multiple parameters and be applicable to a range of compounds. For instance, a saturated solution prepared to determine aqueous solubility may subsequently be used to determine partition coefficient, by addition of *n*-octanol.

Note at this stage that determination of *approximate* values is acceptable in order to make a go/no go decision in respect of a particular candidate and so assays do not need to be as rigorously validated as they do later in formulation development. Table 1.3 lists a range of molecular properties to be measured during preformulation, in chronological order, and the assays that may be used to quantify them. These properties are a function of

Table 1.4   Macroscopic (bulk) sample properties and the techniques used to determine them.

Derived property	Technique
Melting point	DSC or melting point apparatus
Enthalpy of fusion (and so ideal solubility)	DSC
Physical forms (polymorphs, pseudopolymorphs or amorphous)	DSC, XRPD, microscopy
Particle shape • Size distribution • Morphology • Rugosity • Habit	Microscopy Particle sizing BET (surface area)
Density • Bulk • Tapped • True	Tapping densitometer
Flow	Angle of repose
Compressibility	Carr's index Hausner ratio
Excipient compatibility	HPLC, DSC

molecular structure. Once known, further macroscopic (or bulk) properties of the drug candidate can be measured (Table 1.4). These properties result from intermolecular interactions. Note also that determination of chemical structure does not appear, as it is assumed that the chemists preparing the candidate molecules would provide this information. Note also that solubility will be dependent upon physical form (polymorph, pseudopolymorph or amorphous).

Full characterisation of a drug substance should be possible with just five techniques: ultraviolet (UV) spectrophotometry, thin-layer chromatography (TLC), high-performance liquid chromatography (HPLC), differential scanning calorimetry (DSC) and dynamic vapour sorption (DVS). This explains the popularity of these techniques in pharmaceutical development laboratories and so their basic principles are outlined below. Other, more specialised techniques (such as X-ray powder diffraction, XPRD) provide additional information. Application of the techniques is discussed in later chapters, but the basic principles are discussed below.

Note that in the limit the sensitivity of the assay will be dependent upon the purity of the sample (greater levels of impurity lowering sensitivity) and so assay development should be undertaken with the purest sample obtainable. Sensitivity can be expressed in many ways, but commonly detection limits (DL) or quantification limits (QL) are specified. There are many ways of calculating DL and QL values. ICH Guideline Q2(R1) (2005) defines the following:

$$DL = \frac{3.3\sigma}{s} \tag{1.1}$$

$$QL = \frac{10\sigma}{s} \tag{1.2}$$

where σ is the standard deviation of the blank measurement and s is the slope of the calibration plot. Since all assays require understanding of concentration terms, these concepts will be discussed first.

1.3   Concentrations

Concentration terms simply define the ratio of two components in a particular sample. The minor component is termed the solute and the major component is termed the solvent. It does not matter what the physical forms of the solute and solvent are (i.e. they can be solid, liquid or gas, although certain combinations are not usually encountered, such as a gas dissolved in a solid).

Importantly, a concentration term specifies the amount of solute present *per unit of solvent*. Thus, defining a concentration gives no information on how large the sample is; everything is normalised to a particular unit. So, for instance, if a sample is defined as a 1 M aqueous solution of aspirin, there is a mole of aspirin in every litre of water. It is not possible to know from this statement how much solution there is. If, instead, the sample was defined as 500 mL of a 1 M aqueous solution of aspirin, there is sufficient information to know everything about what material is present and in what quantity.

1.3.1   Units of concentration

The amounts of solute and solvent can be specified a number of ways. The most commonly encountered units in pharmaceutics are:

- Molar (M, moles per litre)
- Molal (m, moles per kg)
- Percentages (w/w, w/v, v/v)
- Weight per volume (mg mL^{-1})
- Parts per million (ppm)
- Mole fraction (x)

Since it is possible to define concentrations with a multitude of terms, care must be taken when comparing systems expressed in different units. The major issue to be accounted for is the molecular mass of the solute.

Example 1.1   Which of the following pairs of solutions (assume there is 1 L of each) contains the same number of solute molecules?

   (a)   2 % w/v formoterol fumarate and 2 % w/v salbutamol sulphate

   (b)   0.1 M formoterol fumarate and 0.1 M salbutamol sulphate

The answer is (b), because the amount of solute is expressed in terms of molarity, which is independent of molecular weight.

    For small organic molecules, such as the majority of drugs, differences in the number of molecules between solutions expressed in weight percentages may be small, but as the molecular weight of the solute increases (where polymeric excipients are used, for instance) the differences can become significant. Care must be taken when constructing and interpreting an experimental series based on percentage concentrations that differences observed between solutes do not arise simply as a result of different numbers of solute molecules per unit volume.

    Molar concentrations avoid this problem and so molar is the standard unit of concentration used in the SI[1] (le Système International d'Unités) nomenclature. If Z is the molecular weight of a solute, then Z grams of that solute contains 1 mole (6.022×10^{23}) of molecules.

    The difference between molar (M) and molal (m) is the same as the difference between % w/v and % w/w (i.e. 1 M is 1 mole per litre while 1 m is 1 mole per kilogram).

    In pharmaceutics the molarities of typical solutions may be very low and hence the most frequently encountered units are those based on weight or volume fractions. Many dosage forms are solids and thus are more amenable to percentage concentration expressions. Also, if the molecular weight of a new drug substance is not known, then it is not possible to calculate molar or molal concentrations.

Example 1.2   What do the following concentration terms mean?

   (a)   0.1% w/v

   (b)   2% w/w

---

[1] Interestingly, three countries have not adopted SI nomenclature, Liberia, Burma and the United States, although as of 2010 Liberia is gradually introducing metric units. The United Kingdom uses an eclectic mix of SI units in science and metric and Imperial units in everyday life.

In the case of (a) the concentration term (w/v) implies a solid solute has been dissolved in a liquid solvent; 0.1% implies that the ratio of solute to solvent is 0.1:100. So 0.1% w/v means 0.1 g of solute in 100 mL of solvent.

In the case of (b) the concentration term (w/w) implies a solid solute has been dissolved in a solid solvent; 2% implies that the ratio of solute to solvent is 2:100. So 2% w/w means 2 g of solute in 100 g of solvent.

Another point to remember is that percentage terms are expressed per 100 mL of solvent while molar terms are expressed per litre of solvent. Although weight percentage terms are common in pharmaceutics, again the low concentrations often used make the numbers small. Also, many medicines are defined as weight of drug per unit dose (50 mg per tablet for instance), so weights per unit volume concentrations are very often used:

- 2 mg mL^{-1}
- 50 mg L^{-1}
- 10 g L^{-1}

Example 1.3   Do the following solutions contain equal numbers of molecules?

(a) 5 mg mL^{-1} paracetamol and 5 mg mL^{-1} ibuprofen

(b) 10 mg mL^{-1} nicatinamide and 10 mg mL^{-1} isonicatinamide

Not in the case of (a) as the molecular weights of the drug substances are different. The only concentration terms that normalise for numbers of molecules are molarity or molality. In the special case (b) the drug substances have the same molecular weight and so the numbers of molecules are equal.

The term ppm is less commonly encountered in pharmaceutics, being more associated with gases or very dilute contaminants in solution; 1 ppm means 1 part of solute to a million parts of solvent (easily remembered as 1 mg per litre).

There is one further way of expressing concentration: mole fraction (x). The mole fraction of a component is defined as the number of moles of that component divided by the total number of moles of all of the components in the system:

$$x_a = \frac{\text{Number of moles of component } a}{\text{Total number of moles of all components in system}} \qquad (1.3)$$

Mole fractions are dimensionless and must always have a value between 0 and 1. The sum of the mole fractions of all the components in a system must equal 1. Mole fraction units are useful if there are two or more solutes in the same solvent.

Example 1.4   A solution for intravenous injection is prepared at 25 °C with the following constituents: water (50 g, RMM 18), lidocaine hydrochloride (1 g, RMM 270.8) and epinephrine (0.5 mg, RMM 183.2). Calculate:

- The mole fraction of lidocaine hydrochloride

- The mole fraction of epinephrine

- The mole fraction of water

Firstly, the number of moles of each component must be calculated:

$$\text{Number of moles of lidocaine hydrochloride} = \frac{1}{270.8} = 0.00369$$

$$\text{Number of moles of epinephrine} = \frac{0.0005}{183.2} = 0.00000273$$

$$\text{Number of moles of water} = \frac{50}{18} = 2.78$$

and so

$$x_{\text{lidocaine HCL}} = \frac{0.00369}{0.00369} + 0.00000273 + 2.78 = 0.00133$$

$$x_{\text{epinephrine}} = \frac{0.00000273}{0.00369} + 0.00000273 + 2.78 = 0.00000098$$

$$x_{\text{water}} = 1 - 0.00133 - 0.00000098 = 0.9987$$

Summary box 1.1

- Concentrations define the amount of solute per unit volume or mass of solvent.

- Molar or molal concentrations can be compared in terms of numbers of solute molecules.

- Percent or weight/volume terms are more common in pharmaceutics and can be converted to molar/molal concentrations if the molecular weight of the solute is known.

- $1\% \text{ w/v} \equiv 10 \text{ g L}^{-1} \equiv 10 \text{ mg mL}^{-1}$

- $1 \text{ ppm} \equiv 1 \text{ mg L}^{-1}$

Table 1.5 UV absorbance maxima for a range of common functional groups (data from Wells (1988)).

Chromophore	λ_{max} (nm)	Molar absorption (ε)
Benzene	184	46 700
Naphthalene	220	112 000
Anthracene	252	199 000
Pyridine	174	80 000
Quinoline	227	37 000
Ethlyene	190	8000
Acetylide	175–180	6000
Ketone	195	1000
Thioketone	205	Strong
Nitrite	160	–
Nitroso	302	100
Nitro	210	Strong
Amino	195	2800
Thiol	195	1400
Halide	208	300

1.4 UV spectrophotometry

Unless there is a good reason not to, the primary assay developed during preformulation will be based on UV spectrophotometry. Many factors contribute to the popularity of the technique, including familiarity, cost, amount of solution used and the fact that the majority of drug substances contain at least one functional group that absorbs in the ultraviolet (UV) region (190–390 nm). Table 1.5 lists the UV absorbance maxima for a series of common functional groups (called *chromophores*).

Since a chromophore is a functional group with absorption in the UV range, excitation of the solute with the appropriate wavelength of light will reduce the amount of light passing through the solution. If the original light intensity is I_0 and the amount of light passing through the sample (the *transmitted light*) is I, then the amount of light absorbed will be a function of the concentration of the solute (C) and the depth of the solution through which the light is passing (the path length, l), usually expressed as the Beer–Lambert equation:

$$\text{Absorbance} = \log \frac{I}{I_0} = \varepsilon Cl \qquad (1.4)$$

where ε is a constant of proportionality called the molar absorption coefficient. Higher values of ε mean greater absorbance by the solute. Values of ε for a range of functional groups are given in Table 1.5; it can be seen that groups containing large numbers of delocalised electrons, such as those

Table 1.6   The effect of auxochromes on the UV absorbance of the parent compound C_6H_5–R (data from Wells (1988)).

Substituent	λ_{max} (nm)	Molar absorption (ε)
–H	203.5	7400
–CH_3	206.5	7000
–Cl	209.5	7400
–OH	210.5	6200
–OCH_3	217	6400
–CN	224	13 000
–COO^-	224	8700
–CO_2H	230	11 600
–NH_2	230	8600
–$NHCOCH_3$	238	10 500
–$COCH_3$	245.5	9800
–NO_2	268.5	7800

containing benzene rings, have much greater ε values than groups containing simple carbon–carbon double bonds. Many drug substances contain aromatic moieties of carbon–carbon double bonds, which is why UV spectroscopy is a good first choice assay.

The absorbance of a chromophore can be affected by the presence of an adjacent functional group if that group has unshared electrons (an *auxochrome*). A list of common auxochromes and their effects on the molar absorption coefficients of their parent benzene ring is given in Table 1.6.

Use of the molar absorption coefficient is fine when concentrations are expressed in molar terms. However, as noted above, in pharmaceutics it is more common to express concentrations in percentage terms, in which case an alternative constant of proportionality must be defined. This is usually of the form of a specific absorption coefficient (A). Although any such constant may be defined, usually the reference value is the absorbance of a 1% w/v solution in a 1 cm path length UV cuvette:

$$\text{Specific absorption coefficient} = A_{1\ cm}^{1\%} \tag{1.5}$$

The Beer–Lambert equation therefore becomes

$$\text{Absorbance} = A = A_{1\ cm}^{1\%}\,Cl \tag{1.6}$$

Assuming a 1 cm path length cuvette is used then

$$C = \frac{A}{A_{1\ cm}^{1\%}} \tag{1.7}$$

Knowledge of the value of $A_{1\ cm}^{1\%}$ allows determination of the concentration of a solution by measurement of its absorbance (and will yield an answer in % w/v). Values of $A_{1\ cm}^{1\%}$ are often quoted in pharmacopoeial monographs.

Example 1.5   What is the concentration of a solution of buflomedil hydrochloride that gives an absorbance reading of 0.5 at 275 nm ($A_{1\ cm}^{1\%}$ = 143 at 275 nm)?
From Equation (1.5),

$$C = \frac{0.5}{143} = 0.0035\%\text{w/v}$$

As might be expected, the molar absorption coefficient and the specific absorption coefficient are related through the molecular mass of the solute.

Study question 1.1   Show that the molar absorption coefficient (ε) and the specific absorption coefficient ($A_{1\ cm}^{1\%}$) are related according to the following relationship:

$$\varepsilon = A_{1\ cm}^{1\%} \times \left(\frac{\text{Molecular weight}}{10}\right)$$

The $A_{1\ cm}^{1\%}$ value for most drug substances lies in the range 200–1000 with a mean of ca. 500. A 10 µg mL^{-1} solution of a drug with $A_{1\ cm}^{1\%}$ = 500 would give an absorbance of 0.5, well within the range of UV spectrophotometers. Compounds with $A_{1\ cm}^{1\%}$ < 50 are generally too poorly absorbing for successful UV analysis.

1.4.1 Method development for UV assays

If the compound has good aqueous solubility then water is the most appropriate solvent. Frequently, however, a drug substance will have poor aqueous solubility, in which case an alternative strategy is required. Methanol is a good solvent as a first attempt, as it is a good solvent for both polar and nonpolar drugs, it does not have significant UV absorbance and solubility is often nearer to ideal (see Chapter 4). An additional benefit is that it is miscible with water, so the drug substance can initially be dissolved in a small volume of methanol and then diluted with water. Dilution is best achieved with 0.1 M HCl or NaOH (as appropriate, depending upon whether the drug substance is a weak acid or base) since this will maximise ionisation, and hence solubility, and neither solute absorbs in the UV region.

Table 1.7   Suitable solvents for UV analysis.

Solvent	UV cut-off wavelength (nm)
Water	200
Methanol	200
Ethanol (96%)	200
Propanol	200
Isopropanol	210
Butanol	210
Pentanol	210
n-Hexane	210
n-Heptane	210
Glycerol	200
Acetonitrile	200
Cyclohexane	210
iso-Octane	210
Dioxane	220
Ethyl ether	220
Chloroform	245

Other solvents may be used in UV spectroscopy but their UV cut-off (the wavelength below which they absorb significantly) may differ from water (Table 1.7). In the worst case a solubilising agent can be added, but care must be taken to ensure it does not absorb in the UV range or, if it does, to correct for the absorbance with a suitable blank.

The performance of the instrument should be checked prior to use with reference standards. (This is true for all analytical instruments. Institutions such as the National Institute for Standards and Technology, NIST, or the Laboratory of the Government Chemist, LGC, can advise on and supply certified reference materials, CRMs). A solution of holmium oxide in perchloric acid solution can be used for wavelength calibration (Weidner *et al.*, 1985) and is specified in the Ph Eur, while a solution of potassium dichromate can be used to check absorbance (Burke and Mavrodineanu, 1976, 1977). The properties of these solutions, as well as typical specifications for a well-performing instrument are given in Table 1.8.

A full spectrum scan of the solution will allow assessment of the most appropriate wavelength (λ) for analysis. Ideally, this would be the wavelength at which maximum absorption is seen (λ_{max}), corresponding to the tip of an absorbance peak. Failing that, alternative options (in order of preference) are an absorbance valley, a shoulder or a slope. The full spectrum should be checked at low and high concentrations to ensure that the λ chosen is not concentration-dependent (for instance, if the peak of an absorbance maximum is chosen, a small shift in the peak with concentration may lead to a significant fall in absorbance). If the value of λ_{max} is close to that of a suspected

Table 1.8 Specifications for a UV spectrometer (data from Wells (1988)).

Attribute	Specification		
Wavelength (λ)	Holmium oxide in perchloric acid solution ± 1 nm between 200 and 400 nm		
Characteristic maxima	241.15, 287.15, 361.50 and 536.30 nm		
Absorbance	Potassium dichromate solution		
	λ	$A_{1\,cm}^{1\%}$	$A_{1\,cm}^{1\%}$ limit
	235 nm	124.5	± 1.7
	257 nm	144	± 1.7
	313 nm	48.6	± 1.7
	350 nm	106.6	± 1.7
Cuvettes	Quartz, path length ± 0.005 cm		
Solvents	Absorbance <0.4 (ideally <0.2) relative to air		
Path length	Usually 1 cm, although 1 mm cuvettes available		
Temperature	$20 \pm 1\,^{\circ}C$		

impurity, or that of a solubilising agent if used, then an alternate wavelength can be chosen.

Preparation of a 10 µg mL^{-1} solution should allow rapid determination of the $A_{1\,cm}^{1\%}$ value. If it is not possible to prepare a solution of exactly 10 µg mL^{-1} then any alternative concentration may be used, since $A_{1\,cm}^{1\%}$ is proportional to concentration. Rearranging Equation (1.7),

$$A_{1\,cm}^{1\%} = \frac{A}{C(\%w/v)} \tag{1.8}$$

If solutions are prepared in mg mL^{-1} units then

$$A_{1\,cm}^{1\%} = \frac{10A}{C(mg\ mL^{-1})} \tag{1.9}$$

If the solute absorbs very strongly that the stock solution must be diluted, then the dilution factor (F) can also be accounted for in determining the value of $A_{1\,cm}^{1\%}$:

$$A_{1\,cm}^{1\%} = \frac{AF}{C(\%w/v)} \tag{1.10}$$

$$A_{1\,cm}^{1\%} = \frac{10AF}{C(mg\ mL^{-1})} \tag{1.11}$$

Alternatively, a full calibration plot can be prepared from a series of stock solutions of known concentrations. Although this takes more time and uses more material, the benefits are that linearity of response is demonstrated as a function of concentration and errors are reduced. Care must be taken that, if the solute degrades, solutions are prepared and analysed as quickly as possible.

Summary box 1.2

- Assay development is the first job of a preformulation scientist.

- UV spectrophotometry is the primary assay of choice and can be used to determine solubility and the partition coefficient.

- Analytes must possess a UV chromophore. The absorbance of a chromophore can be affected by an adjacent auxochrome.

- Absorbance strength is quantified by the molar absorption coefficient, ε, or the specific absorbance, $A_{1\,cm}^{1\%}$.

- Water or methanol are first choice solvents for UV assays.

1.5   Thin-layer chromatography (TLC)

TLC derives from paper chromatography, an early method used to separate bands of colours from botanical samples (hence the name chromatography). In paper chromatography a solvent containing the solutes to be separated advances up the paper substrate by capillary action. The solvent advances at a particular rate but the solutes repeatedly absorb and desorb from the cellulose fibres. The varied absorption and desorption characteristics of the solutes means that their rates of progression will be different, and always slower than that of the solvent front. Hence, with time, the solutes will become separated.[2] The solvent is termed the *mobile* phase and the paper the *stationary* phase.

The principle of TLC is the same, but the stationary phase is deposited on to an aluminium or glass substrate (and so does not have to be cellulose). Typical stationary phases include silica, derivatised silica, alumina and cellulose/cellulose derivatives (in which case TLC becomes analogous to paper chromatography). Thus the technique is much more flexible than the paper method and can be optimised to maximise separation of a given mixture of solutes. Although considered an older, and less relevant, technique than HPLC (see below) the development of stationary phases for HPLC has led

---

[2] Any parent whose child has dropped a sugar-coated chocolate sweet on to their clothing and subsequently dribbled will have seen this effect in action.

to an improved range of stationary phases for TLC, and thus some resurgence in its use. TLC can also be used as a rapid screen for selection of HPLC solid and stationary phases.

In general, spots of sample are placed near to the bottom edge of the TLC plate, often on a line marked with a pencil. The marked plate is then stood vertically in an enclosed glass chamber that has a small volume of solvent (mobile phase) at the bottom. The mobile phase advances up the plate, carrying the sample chemicals with it. The distance that the solvent front has advanced, relative to the pencil line, is recorded (d). The final position of each chemical species (d_s) will be a function of its affinity for the stationary phase relative to its affinity for the mobile phase and will be less than d. The resolution factor (R_f) for each chemical can then be defined as;

$$R_f = \frac{d_s}{d}$$
(1.12)

Unless the chemicals are coloured, the plate will have to be 'developed' by staining in order to measure the advancement of each component. Alternatively, a UV light may cause the chemicals to fluoresce.

1.5.1 TLC method development

1.5.1.1 Preparation of the plates Both normal and reverse-phase TLC plates are commercially available and so do not need to be prepared by the analyst unless a special stationary phase is required. A 5 mm band of the coating material is removed from each vertical side to prevent 'edge effects', which may distort the solvent front.

1.5.1.2 Spotting Samples are extracted into chloroform (or a mixture of chloroform and methanol as appropriate) and a small volume (10–60 μL) is 'spotted' (using a glass capillary) on to the starting line of the plate (Figure 1.1). The spots are usually made about 20 mm from the base of the plate and each spot is approximately 20 mm from its neighbour. This means that nine samples can be run simultaneously on a standard 200 mm square plate. The spots are either air-dried or blow-dried prior to immersion of the base of the plate in the mobile phase in order to remove any interference that may be caused by the presence of the solvent used for spotting.

1.5.1.3 Separation A 10 mm layer of the mobile phase is poured into the base of the closed tank. The spotted TLC plate is then stood vertically in the liquid and the solvent allowed to move up the plate. The solvent front should rise a minimum of 150 mm before development of the plate.

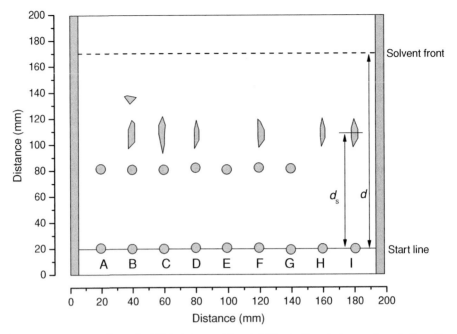

Figure 1.1   Layout of a typical TLC plate used for stability testing during preformulation. Key: A, 2% impurity standard; B, 50 °C stability sample; C, 1% impurity in laboratory reference sample of drug; D, 37 °C stability sample; E, 1% impurity; F, 30 °C stability sample; G, 0.5% impurity; H, 4 °C stability sample; I, Laboratory reference standard (redrawn from Wells (1988), with permission from John Wiley & Sons, Inc.).

1.5.1.4   Mobile and stationary phase selection   Silica gel and alumina are the two most commonly used stationary phases. Silica gel is the default option as it allows separation of compounds with a wide range of polarities. If the compound exhibits very low polarity then retention on silica gel is poor and if the compound is highly polar retention on silica gel is too high; in either case reverse-phase TLC can be employed. Alumina has a more complex surface chemistry than silica gel, with hydroxyl groups, aluminium cations and oxide anions and its pH and degree of hydration can significantly affect separation. A range of chemically derivatised silica substrates is available (Table 1.9), which are used in reverse-phase TLC.

Table 1.9   Derivatised silica stationary phases used in reverse-phase TLC.

Derivatised silica	Functional group	Application
Alkylsiloxane	$Si-CH_3$	Reverse-phase TLC. Separation of:
	$Si-C_2H_5$	• Water-soluble polar compounds
	$Si-C_8H_{17}$	• Weak acids and bases
	$Si-C_{18}H_{37}$	• Strong acids and bases

Table 1.10   Solvent strengths in normal- and reverse-phase TLC.

Normal-phase (silica gel)	Elution strength on SiO_2	Solvent	Reverse-phase (C_{18} derivatised silica)
Weakest	0.01	Hexane	Strongest
	0.04	Cyclohexane	
	0.12	CCl_4	
	0.23	Toluene	
	0.25	Benzene	
	0.32	Dichloromethane	
	0.35	Tetrahydrofuran	
	0.38	Ethylene dichloride	
	0.39	Methyl ethyl ketone	
	0.46	Methyl acetate	
	0.47–0.53	Acetone	
	0.5	Acetonitrile	
	0.68	Ethanol	
	0.73	Methanol	
	Large	Acetic acid	
Strongest	Large	Water	Weakest

Selection of the mobile phase is based on the concepts of *strength* and *selectivity*. Strength is the ability of a solvent to cause solute migration and is a composite property of the solvent and stationary phase, not an intrinsic property of the solvent alone. For example, water is a strong solvent when silica gel is the stationary phase but a weak solvent when a derivatised C_{18} silica layer is the stationary phase (Table 1.10). Selectivity is the ability of a solvent to separate solutes.

The stationary phase is thus selected first, based on the likely polarity of the sample, and the mobile phase is selected in response, firstly on strength and then on selectivity. The strength of a mobile phase may be adjusted by addition of a co-solvent. For normal-phase TLC, hexane is used as the strength-adjusting solvent for weakly and moderately polar compounds. For polar compounds, the strongest solvent that fails to cause migration of the solute is used. For reverse-phase TLC water is always the strength-adjusting solvent so water miscible solvents must be used during optimization of the method. Poole and Dias (2000) list the first choice solvents for normal and reverse phase TLC (Table 1.11) and provide an excellent guide to TLC method development.

1.5.2   High-performance TLC

The continued development of TLC has led to improvements both in terms of instrumentation (for instance, to enable more precise spotting of samples on to the plate or to improve quantitative evaluation of the movement

Table 1.11 Mobile-phase solvents for TLC method development (data from Poole and Dias (2000)).

Normal phase	Reverse phase
1. Hexane	1. Methanol
2. Toluene	2. 2-Propanol
3. Methyl-*t*-butyl ether	3. 2,2,2-Trifluroethanol
4. Dichloromethane	4. Acetone
5. Chloroform	5. Pyridine
6. Ethyl acetate	6. Acetonitrile
7. Acetone	
8. Pyridine	
9. Triethylamine	
10. Acetonitrile	
11. Methanol	
12. 2-Propanol	
13. 2,2,2-Trifluroethanol	
14. Acetic acid	
15. Water	

of the samples) and in the particle size of the stationary phase. The silica gel used in TLC typically has a particle size range of 10–60 μm, but high-performance TLC (HP-TLC) utilises ultrafine particles smaller than 5 μm. In addition, smaller plates are used (often 10 cm × 10 cm) and sample volumes are reduced by an order of magnitude. One problem with HP-TLC is that the rate of advancement of the solvent front is often seen experimentally to reduce after a short distance. For this reason, HP-TLC methods have been developed that hold the TLC plate under pressure (forced-flow TLC).

Summary box 1.3

- TLC separates mixtures of compounds based on their affinity for the stationary phase relative to their affinity for the mobile phase.

- TLC can be normal-phase or reverse-phase.

- The stationary phase is selected first – silica gel will separate most compounds unless polarity is very high or very low, in which case a reverse-phase solid substrate is used.

- The mobile phase is selected after the stationary phase on strength (ability to cause migration of solutes) and selectivity (ability to separate solutes).

- TLC can be used to shorten HPLC method development times.

1.6 High-performance liquid chromatography

HPLC is probably the most common analytical technique used in pharmaceutical development. Its widespread application is a direct result of its ability to separate a vast range of compounds; in other words, it can be applied to virtually any sample. Note (as is always the case with chromatographic methods) that chromatography in itself is not an assay; chromatography separates mixtures of analytes and a downstream detector performs the assay. In the case of HPLC, the detector is typically UV, fluorescent, mass spectrometric, flame-ionisation or refractive index (RI).

The analyte mixture is injected into a mobile phase, which is flowed through a column at high pressure, typically up to 6000 psi (43 MPa) although ultra-high-performance (UHPLC) instruments can operate up to 18 000 psi (130 MPa), containing a solid (stationary) phase. As with TLC, different analytes interact with the stationary phase to different extents and hence the time taken for elution (the retention time, t_r) is different and characteristic for each analyte. Retention times (which can be altered for each analyte through selection of mobile and stationary phases) will ultimately reduce to the balance of the affinity of the solute for the stationary phase relative to its affinity for the mobile phase. A detector on the eluent stream of the column detects the analytes as they elute. Ideally, each analyte will elute as a separate (resolved) peak. Unlike TLC, where the analytes remain on the plate during development and analysis, HPLC requires all analytes to be eluted from the stationary phase to be detected. Where there is a particularly strong affinity for an analyte to the stationary phase, removal may be impossible and the column is considered 'poisoned'.

The most important part of any HPLC system is the column. The column consists of three parts: the hardware, the matrix and the stationary phase. The hardware is the metal (usually) housing in which the matrix and stationary phase are placed. The matrix is a solid substrate to which the stationary phase is chemically bound. Typically, silica gel beads are used as the matrix material, because the hydroxyl groups on the surface are readily chemically modified, they are easily manufactured to a consistent particle size and they do not compress under the (significant) pressure of the mobile phase (although they can dissolve at higher pH values). The physical nature of the matrix is often the biggest factor in the separation efficiency of the column; smaller matrix particle sizes result in greater interaction with the analytes and hence greater separation efficiency (an increase in the number of theoretical plates). However, the use of smaller particle sizes also results in increased backpressure and a higher risk of column blockage. Hence, 5 Å columns are more frequently used in development of assays than 3 Å columns, although once an assay has been developed 3 Å columns can be employed. Particle size distribution also affects separation efficiency and it may well be observed

Table 1.12   Stationary phases used in reverse-phase HPLC.

Stationary phase	Application
C_3, C_4, C_5 (propyl, butyl, pentyl)	Ion-pairing (peptides with hydrophobic residues and other large molecules)
C_8 (octyl or MOS)	Wide application. Especially good for nucleosides, steroids, peptides and small hydrophilic proteins
C_{18} (octadecyl or ODS)	Wide application for nonpolar analytes (up to pH 7 mobile phase)
C_{18} Xterra or Zorbax-extend	Designed to tolerate higher pH (up to pH 11) mobile phases
Phenyl	Selective for alkyl phases (resolution of aromatic analytes)
Nitrile (CN or cyano)	Polar; will increase retention of polar analytes

that 'equivalent' columns from different manufacturers will result in different retention times. Note also that the solid-phase particles may have a degree of porosity, which will affect retention times differently depending upon the molecular weight of the analyte species.

1.6.1   Normal- and reverse-phase HPLC

In normal-phase the stationary phase is polar and the mobile phase is non-polar so polar compounds are retained longer on normal-phase columns than nonpolar compounds. Typical stationary phases include underivatised silica, nitrile, amino, glycerol and nitro gels. Chiral separation is usually achieved on normal-phase columns.

In reverse-phase HPLC the stationary phase is nonpolar and the mobile phase is polar and so nonpolar compounds are retained longer than polar compounds. Reverse-phase HPLC is most commonly employed to separate pharmaceutical analytes. Common stationary phases are C_4 (butyl), C_8 (octyl), C_{18} (ODS) and phenyl. In general, longer alkyl chains, higher phase loading and higher carbon loads cause greater retention of nonpolar analytes. Typical applications of reverse-phase stationary phases are given in Table 1.12.

The mobile phase is also important when designing an HPLC assay, for although the column has the greatest effect on resolution, mobile-phase selection is more easily varied by the user and can be used to adjust retention times, both in absolute and relative terms. Separation of closely eluting isomers, for instance, may be achieved with subtle modifications to the mobile phase, such as a change in counterion. Usually, in reverse-phase HPLC the mobile phase consists of an aqueous buffer mixed with a non-UV active water miscible organic solvent.

The pH of the aqueous phase is chosen to control the degree of ionisation of the analytes. Ionised analytes will elute more quickly, which tends to lead

Table 1.13   Buffer solutions for reverse-phase HPLC.

Buffer	pK_a	pH range	UV cut-off
H_3PO_4/KH_2PO_4	2.12	1.1–3.1	<200 nm
KOAc/AcOH	4.8	3.8–5.8	210 nm
KH_2PO_4/K_2HPO_4	7.21	6.2–8.2	<200 nm
NH_4OH	9.2	8.2–10.2	200 nm
$Et_3NH/Et_3NH.HCl$	11.0	10–12	<200 nm

to a better peak shape, while un-ionised analytes will have longer retention times. Typically a 10–50 mM solution of aqueous buffer is used (the most common being 17 mM, or 0.085% w/v, H_3PO_4), listed in Table 1.13, although a volatile buffer is beneficial where mass spectroscopy is the detection system.

The selection of the organic phase plays an important role in the separation of analytes, since the solubilities of compounds will vary independently in different solvents. Organic HPLC solvents are graded in terms of their strength in the same way as noted for TLC. Typical solvents used for reverse-phase HPLC are methanol, acetonitrile and THF. Acetonitrile in particular is very commonly used, as it provides good resolution for a wide variety of compounds. Methanol will form hydrogen bonds with analytes and hence can increase resolution. THF can be difficult fully to remove from the column as it is highly hydrophobic.

1.6.1.1   Solvent gradients   In normal-phase HPLC *isocratic* (one solvent) methods are used, whereas in reverse-phase HPLC solvent *gradients* are employed. This means the ratio of aqueous buffer to organic solvent is varied throughout the run, typically with an increasing organic solvent ratio over time. This will elute polar compounds before nonpolar compounds. Before the column can be used for a further experiment, it must be conditioned with the initial mobile phase, typically for 5–10 min. A common gradient would be from 10 to 90% organic solvent over 15 min. These times would reduce where a UHPLC system is used.

1.6.2   HPLC method development

Numerous factors contribute to the retention time of an analyte and as such HPLC method development is neither quick nor simple (and forms the basis of a book in itself). Often a literature search will result in a method that has been developed for a similar analyte. Alternatively, as noted above, TLC can be used rapidly to select mobile and stationary phases, although care must be taken in extrapolating the solvent system. Many of the nonpolar solvents used in TLC absorb UV, which is the usual detection method following HPLC separation. HPLC also usually requires less polar solvents; halving the content of the most polar solvent used in the TLC is a useful starting point in the development of an HPLC method.

Other factors to consider include retention times, flow rates, column pressure and column temperature, since there needs to be a balance between experiment speed and resolution of analytes. For the most accurate analysis the peak for each analyte needs to be fully resolved (i.e. there should be no overlap with other analyte peaks). This can be achieved with slow flow rates, but at the expense of increased experimental runtime, or with gradient systems. Conversely, integration of peaks is more accurate when they are narrower (as the peak maximum will be higher) and slow flow rates can cause peak broadening.

Choice of mobile phase is critical and often driven first by solubility and then by polarity. A strong solvent should be used for initial development to ensure that all the analytes are eluted from the column and that poisoning, as mentioned earlier, does not occur. The strength can then be reduced to delay elution of components and aid separation. Ahuja and Rasmussen (2007) provide a practical guide to HPLC method development for pharmaceuticals.

Summary box 1.4

- HPLC separates mixtures of compounds based on their affinity for the stationary phase relative to their affinity for the mobile phase.

- Typically, reverse-phase HPLC is used, where the stationary phase is nonpolar and the mobile phase is polar (usually a mixture of an aqueous buffer and a water-miscible organic solvent such as acetonitrile).

- Initially a strong mobile phase is used to ensure all the solutes are eluted, but as a method is developed a solvent gradient is usually employed to fine-tune the retention times of each solute, ensuring good resolution with an acceptable run-time.

1.7 Differential scanning calorimetry

Differential scanning calorimetry (DSC) is one of a group of techniques collectively termed as thermal analysis. In DSC the power required to heat (usually, although samples can also be cooled) a sample in accordance with a user-defined temperature programme is recorded, relative to an inert reference. The heating rate (β) can be linear or modulated by some mathematical function. DSC data are presented either as power versus temperature or heat capacity versus temperature (obtained by dividing the power data by the heating rate). There are two principal designs of DSC, which differ in how the samples are heated.

> *Heat-flux DSC.* In heat-flux DSC a *common furnace* heats the sample and reference materials and the temperature difference (ΔT) between them is recorded. The power change occurring in the sample is directly proportional to the temperature difference.
>
> *Power compensation DSC.* In power-compensation DSC *separate furnaces* heat the sample and reference materials. The instrument varies the power supplied by the two furnaces to maintain the temperature difference between the sample and reference at zero. The power difference (ΔP) between the sample and reference is thus measured directly.

1.7.1   Interpreting DSC data

The DSC signal (power) comprises contributions from two sources: heat capacity (C_p) effects and any other processes (phase transformations or chemical reactions) that the sample might undergo (represented by the generic term $f(T, t)$):

$$\frac{dq}{dt} = C_p \frac{dT}{dt} + f(T, t) \tag{1.13}$$

Thus, if the sample undergoes a phase change or chemical process, or there is a change in heat capacity, there will be a concomitant event in the DSC data.

1.7.1.1   Effect of heat capacity

Temperature and heat are different thermodynamic quantities. Temperature is an intensive property (it does not depend on sample size) while heat is an extensive property (it is directly proportional to sample size). Thus two materials at the same temperature may contain different quantities of heat. Inspection of the data for indium and benzoic acid in Table 1.14 reveals that the former has a higher melting temperature but a lower heat of fusion. Adding heat to a material will increase the rate of molecular movement and so result in an increase in temperature; how much the temperature rises for input of a given quantity of heat will depend on how the energy is utilised in terms of molecular movements (vibrations, rotations and translations). This constant of proportionality is called the heat capacity (C) of a sample (usually quoted as a *specific* heat capacity – the amount of heat required to raise 1 g or 1 mol of a sample by 1 K. Also, only changes in heat capacity can be determined and measurements are made at constant pressure, ΔC_P). Since the decrease in molecular movement if heat

Table 1.14 Certified reference materials (CRM) for calibration of DSC instruments (data from Haines (2002)).

CRM	Melting temperature (°C)	Enthalpy of fusion (J g^{-1})
Cyclopentane	−93.4	8.63
Gallium	29.8	79.9
Benzoic acid	123.0	148.0
Indium	156.6	28.6
Tin	231.9	60.4
Aluminium	660.3	398.0

is removed should be equal but opposite to the increase if heat is added, heat capacity changes can be considered *reversible*.

From the perspective of DSC, if the sample and reference materials have the same heat capacities then they will require the same power to heat them in temperature and as a consequence ΔP will be zero if the sample is not undergoing any other process. Hence the DSC thermal trace will be a horizontal baseline centred at $\Delta P = 0$ (Figure 1.2).

If the heat capacity of the sample is different from that of the reference then different amounts of power will be required to heat both materials at the same rate. In this case ΔP will still appear as a horizontal baseline but it will be displaced from zero (Figure 1.2). Clearly, the extent of displacement

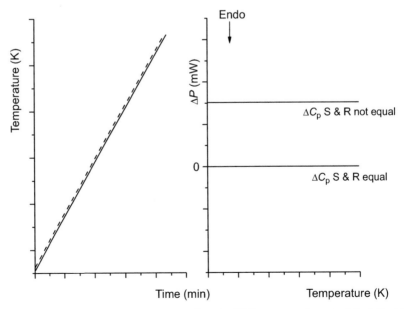

Figure 1.2 Effect of heat capacity on DSC data (solid line – sample material, dotted line – reference material).

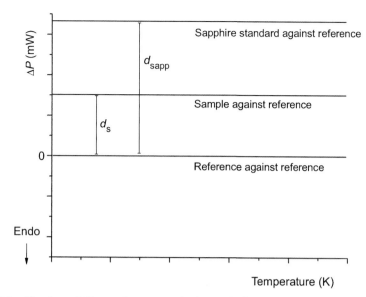

Figure 1.3   The three DSC experiments required to calculate the heat capacity of a sample material.

is proportional to the difference in heat capacity between the sample and reference (indeed, comparison of the displacement with the displacement observed for an inert material of known heat capacity (such as sapphire) allows the heat capacity of a sample to be quantified; Figure 1.3):

$$C_{p,\text{sample}} = \frac{d_s}{d_{\text{sapp}}} C_{p,\text{ sapp}} \tag{1.14}$$

where d represents the displacement from a zero (reference versus reference) baseline and the subscripts 's' and 'sapp' represent sample and sapphire respectively.

1.7.1.2   Effect of phase transitions   If the sample undergoes a phase transition then a deviation from baseline (often a positive or negative peak) will be seen. Melting is a convenient phase transition with which to explain the principle. Before melting, the sample and reference are being heated at the same rate and the power being supplied to both (this discussion is based on using a power compensation instrument but the case for a heat-flux instrument is analogous) is the same (assuming the sample and reference materials are present in equal quantity and have the same heat capacities). Therefore ΔP is zero and a baseline is plotted. When melting commences any power supplied to the sample is used to disrupt intramolecular interactions

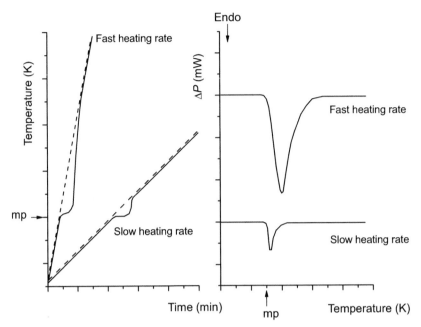

Figure 1.4   Effect of heating rate on DSC data (solid line – sample material, dotted line – reference material).

(endothermic), rather than increase molecular movements, and so the temperature stays momentarily constant. The instrument detects a temperature difference between the sample and reference and so acts to drive ΔT back to zero (in this case by increasing the power supplied to the sample side). As a consequence ΔP is no longer zero and a peak is seen on the thermal trace. Once the melt is complete the sample temperature equalises to that of the reference and a baseline is seen again (Figure 1.4).

Several consequences arise from this discussion. One is that if the heat capacity of the sample changes in going through the phase transition (particularly likely in the event of melting) then the baselines before and after the peak will be different. Another is that the data are scan-rate dependent. The faster the scan rate, the greater the difference in temperature between sample and reference before the instrument responds to ΔT not being zero and the greater the amount of power needed to drive the system back to equilibrium. This effect is also shown in Figure 1.4. A third is that many phase changes (such as crystallisation or loss of water of hydration) or chemical processes can be *irreversible*.

The result is that data recorded at fast heating rates show high sensitivity but poor resolution (because the peaks are broader, so events that occur at similar temperatures can overlap). Conversely, with slow scan rates the data show poor sensitivity but good resolution. Selection of the proper scan

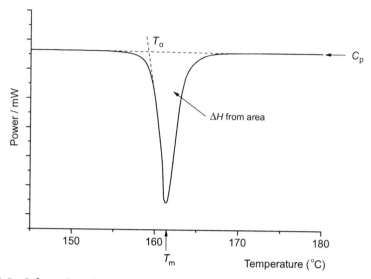

Figure 1.5 Information that can be derived from DSC data (in this case, melting of indomethacin).

rate for an experiment is therefore critical and it must be ensured that any DSC is calibrated at the same scan rate at which any subsequent experiments are run.

Several parameters may be determined from DSC data, including the change in enthalpy (ΔH), the transition temperature (usually determined as an extrapolated onset, T_o, but the peak maximum can also be used, T_m) and the change in heat capacity (ΔC_p) (Figure 1.5).

1.7.2 Modulated-temperature DSC

Reading, Luget and Wilson (1994) first described the principles of modulated-temperature DSC (MTDSC) and an excellent summary of its principles and pharmaceutical applications has recently been published (Reading et al., 2007). Briefly, the linear heating rate (termed the *under-lying* heating rate) is modulated by a periodic function. The modulation can have any form but is typically sinusoidal, square or sawtooth, depending upon the manufacturer of the instrument. In the case of sinusoidal modulation (Figure 1.6), the dependence of temperature with time is described by

$$T = T_0 + \beta t + A_T \sin \omega t \qquad (1.15)$$

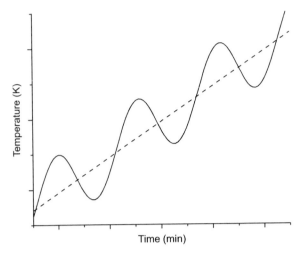

Figure 1.6   Sample temperature as a function of time for MTDSC (solid line, calculated using Equation (1.15)) and the corresponding underlying linear heating rate (dotted line).

where T_0 is the initial temperature, β is the heating rate, ω is the frequency of the modulation and A_T is the amplitude of the modulation. Equation (1.13) can be rewritten to account for the modulation:

$$\frac{dq}{dt} = C_p(\beta + A_T\omega \, \cos(\omega t)) + f'(t, T) + C \, \sin(\omega t) \qquad (1.16)$$

where $f'(t, T)$ is the contribution to the heat flow of any phase transitions or chemical processes (the kinetic response) once the effect of modulation has been removed and C is the amplitude of any kinetic response to the modulation (which is assumed for the purpose of this discussion to be negligible, although in practice a correction factor is often applied).

One component, $C_p(\beta + A_T\omega \cos(\omega t))$, is dependent upon heat capacity effects (which, as discussed earlier, should be reversible, and so is termed the *reversing* heat flow). Heat capacity effects can be considered to occur instantaneously (except for glass transitions); since this term follows a cosine function it should therefore be 0° out of phase with the modulation in the heating rate (assuming endothermic events are plotted up – if plotted down there will be a 180° phase lag).

The second term, $f'(t, T) + C \sin(\omega t)$, is dependent upon a kinetic response (i.e. any process that the sample undergoes, which may be a physical transformation or a chemical reaction that takes a small, but finite, time). Many processes might contribute to the kinetic response (for instance, melting, crystallisation, polymorph transformation, etc.). Some of these processes are reversible (melting) and some can be considered irreversible (transformation to a more stable polymorph, for instance, assuming monotropic polymorphism), but over the time and temperature scales of a typical MTDSC experiment all of these events can be considered to proceed in the forward direction only and so are termed *nonreversing*. Since the kinetic response follows a sine function it should be 90° out of phase with the heat capacity term.

The utility of MTDSC comes in being able to deconvolute the overall heat-flow signal into these two components. Deconvolution requires calculation of the underlying heat-flow signal (which is the average of the modulated response – equivalent to the heat-flow signal that would be recorded for a normal DSC experiment performed at the same underlying linear heating rate). This can be considered equal to

$$\text{Underlying heat flow} = C_p \beta + f'(t, T) \qquad (1.17)$$

Heat capacity reflects the rise in temperature for a given input of heat and so can be determined by comparing the amplitude of the modulated heat flow (A_{mhf}) with the amplitude of the modulated heating rate (A_{mhr}):

$$C_p = \frac{A_{mhf}}{A_{mhr}} \qquad (1.18)$$

The amplitudes are determined using a Fourier transform. From Equations (1.18) and (1.17), it can be seen that the heat capacity (reversing) component of the underlying heat flow is given by

$$\text{Reversing heat flow} = \beta \frac{A_{mhf}}{A_{mhr}} \qquad (1.19)$$

and so the kinetic response (nonreversing) component of the heat flow can be determined by the difference

$$\text{Nonreversing heat flow} = \text{Underlying heat flow} - \beta \frac{A_{mhf}}{A_{mhr}} \qquad (1.20)$$

Heat-flow data recorded with MTDSC thus allow separation of processes into reversing or nonreversing events. This aids both identification and isolation (if multiple events occur at the same temperature) of processes, although being a mathematical routine it is possible to introduce artefacts into the reversing and nonreversing data if the modulation parameters are not selected carefully (see the following section).

1.7.3 DSC method development

DSC is very widely used because, as already noted, there are no specific requirements for a sample, save that it (or a representative fraction of it) can be satisfactorily contained with a DSC pan and that it does not degrade or decompose to a significant extent with heat or upon melting. Method development thus reduces to instrument selection, pan type and experimental parameters.

1.7.3.1 Instrument selection From the user's perspective, whether the DSC is a heat-flux or power-compensation design is essentially irrelevant, since ultimately the data produced will be equivalent. However, by virtue of the fact that heat-flux instruments have only one furnace it tends to be larger than the furnaces found in power-compensation instruments. This means that the maximum heating rates achievable with heat-flux instruments (currently around 200 °C min^{-1}) are slower than those achievable with power-compensation instruments (currently up to 750 °C min^{-1} with commercial instruments – higher heating rates are attainable with research instruments). This seems counterintuitive, but heat-flux instruments must surround the sample and reference pans and sensors with a larger metal heat-sink (to ensure a uniform temperature and to conduct heat from the furnace to the sample and reference) and so there is a greater thermal mass to be raised in temperature.

1.7.3.2 Pan type Pans are usually made of metal (typically aluminium, stainless steel or gold) and have a base and a lid. The seal between the pan and lid can be hermetic (air-tight) or nonhermetic. The choice is important. If the sample contains water that will evaporate into the headspace of the pan, then different responses will be seen in hermetic and nonhermetic pans (the headspace of a hermetic pan will become saturated and evaporation will stop). Similarly, if there is a large expansion in volume of a sample (as it melts for instance) the increase in pressure inside a hermetic pan can cause the seal to fail (often explosively). For this reason, some pans have a pinhole in the lid to prevent a build-up of pressure. In any case, the sample and reference pans should be as closely matched in weight as possible.

1.7.3.3 Experimental parameters The most important decision is the heating rate. Typically, DSC heating rates vary between 2 and 20 °C min^{-1}, although faster heating rates have become popular for physical form characterisation. Since the magnitude and shape of the DSC data will vary with heating rate (Figure 1.4) it is extremely important that the instrument is calibrated both for temperature and for heat flow with a certified reference material (CRM). Typically, CRMs for DSC are highly pure materials with well-established melting points and heats of fusion. The International Union for Pure and Applied Chemistry (IUPAC) recommends a number of CRMs for DSC calibration (Table 1.14). Of these, indium is the most widely used material, although calibration with at least two CRMs is advised.

Selection of heating rate can be an important tool in helping to assign thermal transitions to sample events. It is a good idea to analyse a sample at a minimum of two heating rates, an order of magnitude apart (so 2 and 20 or 20 and 200 $°C$ min^{-1}, for instance). It is also advisable to cool the sample down after the first heating run and then reheat it using the same parameters. This will identify events that are thermally reversible. The applications of these concepts are shown in later sections on physical form characterisation.

Most DSC instruments purge the air space around the sample and reference pans with a flowing gas (typically nitrogen or helium). This flowing gas serves many functions. Firstly, it should be dry (often an in-line desiccant is used) so that if the instrument is operated at subambient temperatures there is no condensation or freezing of water. It also serves as a heat-transfer medium to ensure the pan and contents are at a temperature as close as possible to the heating block (if greater heat transfer is needed, helium is used as the purge gas). Finally, should any gaseous degradants be emitted from the pan, the purge gas ensures they are carried out to waste and do not condense on the instrument.

MTDSC requires selection of additional experimental parameters (underlying heating rate and frequency and amplitude of oscillation). Proper selection is vital to ensure artefacts are not introduced to the data post-deconvolution. In particular, MTDSC assumes that the response of the sample varies linearly with the modulation in temperature. It also assumes that any changes in the underlying heat flow are slow relative to the time scale of the modulation (this allows averaging of the data, required to recover the underlying heat flow). This means there must be many modulations over the course of a transition (a minimum of six is usually indicated; Aubuchon and Gill, 1997). If these conditions cannot be met then deconvolution cannot be achieved. Melting of a pure material is an example where deconvolution usually fails, because as a material melts its temperature will not rise until melting has finished; thus, during melting the temperature of the sample cannot be modulated. Selection of the modulation parameters therefore requires some prior knowledge of the transitions through which the sample will progress and it may be that several experiments will need to be performed with varying parameters to optimize the data. Typical starting values would be an underlying heating rate of 2 $°C$ min^{-1}, frequency 30–60 s and amplitude 0.5–1 $°C$.

Summary box 1.5

- DSC measures the power required to heat a sample relative to that of an inert reference.

- The sample does not need to possess any specific qualities and so the technique has near-universal applicability.

- DSC will detect phase transitions and events that occur with a change in heat capacity.

- Interpretation and assignment of events is aided by changing heating rate and/or heating a sample a second time.

- Modulated-temperature DSC allows separation of heat capacity (reversing) events from kinetic (nonreversing) events.

1.8   Dynamic vapour sorption

In DVS, also known as gravimetric vapour sorption (GVS), the mass of a sample is measured as a function of relative humidity (RH). The mass measurement is typically made with a null-adjust balance. The balance is usually of a modified electronic microbalance design, wherein a lightweight arm is pivoted about an electric coil suspended in a magnetic field. The position of the arm is measured with an optical sensor and any movement induces a current in the coil, which results in it being restored to its equilibrium, or null, position. The sample is counterbalanced either by sending a current to the coil or with a counterweight. One important aspect of this type of balance, apart from its inherent sensitivity, is that the position of the sample in the instrument does not alter if the sample gains or loses mass. The whole apparatus is contained within a thermostat, holding the temperature constant, while the relative humidity can be programmed to increase in discrete steps, at a controlled rate or in response to a change in mass of the sample. The purge gas necessary in DSC is also employed in DVS, but in this case it also acts as the medium via which relative humidity or relative vapour pressure is controlled. Relative humidity can be checked with the use of deliquescent salts, which will uptake a large quantity of water at a specific RH.

1.8.1   DVS method development

As in the case of DSC, any sample can be investigated with DVS. The only experimental variables are selection of a vapour (usually water is used, but ethanol or acetone are common) and the incremental step changes in the desired RH (usually 5–10% steps or a linear ramp).

Summary box 1.6

- DVS (or GVS) measures the mass of a sample as a function of RH or RVP at a constant temperature.

- The sample does not need to possess any specific qualities and so the technique has near-universal applicability.

- The balance is based on a null-adjust design, so the position of the sample remains constant.

- The RH or RVP can be varied in discrete steps or in a linear ramp.

1.9 Summary

Assay development is the first step of any preformulation screen. No physico-chemical parameter can be determined without an assay. Many analytical tools are available but it must be remembered that during preformulation only small amounts of a drug candidate may be available. Hence, initial assays should be widely applicable, use a minimum amount of sample and (prefer-ably) allow determination of several parameters. UV spectrophotometry is an excellent initial choice, as most drugs contain a suitable chromophore and it can be used to determine solubility and the partition coefficient. Chro-matographies (TLC and HPLC) separate mixtures of analytes and so demon-strate purity of the drug substance. They also underpin many subsequent stability assays. Characterisation of physical form can be made with DSC, which can study virtually any solid sample while DVS provides complemen-tary information on stability with respect to humidity.

References

Ahuja, S. and Rasmussen, H. (2007) *HPLC Method Development for Pharmaceuticals*, Separation Science and Technology Series, vol. 8, Academic Press, London. ISBN 0-1237-0540-1.

Aubuchon, S.R. and Gill, P.S. (1997) The utility of phase correction in modulated DSC. *J. Therm. Anal. Cal.*, **49**, 1039–1044.

Burke, R.W. and Mavrodineanu, R. (1976) Acidic potassium dichromate solutions as ultraviolet absorbance standards. *J. Res. Natl Bur. Stand. A*, **80A**, 631–636.

Burke, R.W. and Mavrodineanu, R. (1977) Standard reference materials: certification and use of acidic potassium dichromate solutions as an ultraviolet absorbance standard – SRM 935, NBS Special Publication 260-54.

Haines, P.J. (ed.) (2002) *Principles of Thermal Analysis and Calorimetry*, Royal Society of Chemistry, Cambridge. ISBN 0-8540-4610-0.

ICH Guideline Q2(R1) (2005) Validation of analytical procedures: text and methodology, International Conference on Harmonisation.

Paul, S.M., Mytelka, D.S., Dunwiddie, C.T. *et al.* (2010) How to improve R&D productivity: the pharmaceutical industry's grand challenge. *Nat. Rev. Drug Disc.*, **9**, 203–214.

Poole, C.F. and Dias, N.C. (2000) Practitioner's guide to method development in thin-layer chromatography. *J. Chromatogr. A.*, **892**, 123–142.

Reading, M., Luget, A. and Wilson, R. (1994) Modulated differential scanning calorimetry. *Thermochim. Acta*, **238**, 295–307.

Reading, M., Craig, D.Q.M., Murphy, J.R. and Kett, V.L. (2007) Modulated temperature differential scanning calorimetry. In *Thermal Analysis of Pharmaceuticals* (eds D.Q.M. Craig and M. Reading), CRC Press, Boca Raton, FL. ISBN 0-8247-5814-5.

Weidner, V.R., Mavrodineanu, R., and Mielenz, K.D. (1985) Spectral transmittance characteristics of holmium oxide in perchloric acid solution. *J. Res. Natl Bur. Stand.*, **90**, 115–125.

Wells, J.I. (1988) *Pharmaceutical Preformulation. The Physicochemical Properties of Drug Substances*, John Wiley & Sons, Ltd, Chichester. ISBN 0-470-21114-8.

Answer to study question

1.1 Since ε is the molar extinction coefficient then

$$\varepsilon = \text{Absorbance of 1 M solution} = \frac{\text{Molecular weight (in g)}}{1\,\text{L}} \equiv \frac{\text{mg}}{\text{mL}}$$

In the case of a 1 mg mL^{-1} solution,

$$\text{Absorbance of 1 mg mL}^{-1}\text{ solution} = \frac{\varepsilon}{\text{Molecular weight}}$$

Since

$$A_{1\,\text{cm}}^{1\%} = \text{Absorbance of 1\% w/v solution} = \text{Absorbance of 10 mg mL}^{-1}$$
solution

then

$$\text{Absorbance of 1 mg mL}^{-1}\text{ solution} = \frac{A_{1\,\text{cm}}^{1\%}}{10}$$

and so

$$\frac{A_{1\,\text{cm}}^{1\%}}{10} = \frac{\varepsilon}{\text{Molecular weight}}$$

Additional study questions

1.1 Convert the concentrations of the following aqueous solutions into % w/v (assume aqueous solutions at 25 °C):

 (i) 6 mg ml^{-1} acebutolol (600 mg in 100 mL = 0.6% w/v)

 (ii) 50 g L^{-1} diclofenac sodium (5 g in 100 ml = 5% w/v)

 (iii) 2000 ppm formoterol fumarate (2000 mg in 1 L = 200 mg in 100 ml = 0.2% w/v)

 (iv) 0.118 M clindamycin (RMM 425) (0.118 × 425 = 50 g L^{-1} = 5% w/v)

 (v) 0.053 M morphine sulphate (RMM 376) (0.053 × 376 = 20 g L^{-1} = 2% w/v)

1.2 For each of the following pairs of solutions, which is the greater concentration?

 (i) 0.2% w/v aspirin and 100 mg mL^{-1} formoterol (100 mg mL^{-1})

 (ii) 50 ppm ibuprofen and 40 mg L^{-1} salbutamol sulphate (50 ppm)

 (iii) 5 g L^{-1} metronidazole and 0.5% w/v paracetamol (they are equal)

 (iv) 0.084 M aspirin (RMM 180) and 0.2% w/v diclofenac sodium (aspirin – 0.084 × 180 = 15 g L^{-1}, 0.2% w/v = 2 g L^{-1})

 (v) 0.075 M cyclizine hydrochloride (RMM 266) and 0.084 M aspirin (RMM 180) (0.075 × 266 = 20 g L^{-1} and greater than 15 g L^{-1}, from above)

2
Ionisation Constants

2.1 Introduction

The number of drug substances is constantly changing and thus so is the proportion of acidic and basic compounds. Wells (1988) estimates that 75% of drugs are weak bases and 20% are weak acids (with the remainder being nonionic or amphoteric species or alcohols) while Manallack (2007) reports an analysis of the 1999 World Drug Index that suggests 62.9% of the collection were ionisable between pH 2 and 12. Selected properties of the 10 best selling drugs worldwide in 2010 are shown in Table 2.1. Note the number of biological drug substances (all monoclonal antibodies), which present a different formulation challenge (usually solubility is high but retention of tertiary structure and activity during processing and formulation is paramount).

Understanding acid and base behaviour is extremely important, not only because of the sheer number of ionisable drug substances but because the solubility of an acidic or basic drug substance will be pH-dependent. Possession of an ionisable functional group also opens the possibility of solubility manipulation via salt formation, so long as the salt is stable at physiological pH.

2.2 Ionisation

Understanding the character of weak acids and bases starts with consideration of the partial ionisation that will occur following dissolution in water. Using the notation HA to represent an acid

$$HA + H_2O \rightleftharpoons H_3O^+ + A^-$$ (2.1)

where H_3O^+ is the hydronium ion (the H^+ ion does not exist in isolation in water). Strong acids will dissociate almost completely, shifting the position of

Essentials of Pharmaceutical Preformulation, First Edition. Simon Gaisford and Mark Saunders.
© 2013 John Wiley & Sons, Ltd. Published 2013 by John Wiley & Sons, Ltd.

Table 2.1 Solubility and pK_a data for the 10 best selling drugs of 2010.

Brand name	Active	pK_a	Solubility
Lipitor	Atorvastatin calcium	4.46	20.4 μg mL^{-1} (pH 2.1)
			1.23 mg mL^{-1} (pH 6.0)
Plavix	Clopidogrel bisulfate	n/a	0.051 mg mL^{-1}
Seretide	Fluticasone and	12.5	0.51 μg mL^{-1}
	salmeterol	9.9	Sparingly soluble
Nexium	Esomeprazole magnesium trihydrate	4.0	0.3 mg mL^{-1}
Seroquel	Quetiapine fumarate	6.8	35 mg mL^{-1} (0.1N HCl)
			1.3 mg mL^{-1} (pH 7.4)
Crestor	Rosuvastatin calcium	4.6	7.8 mg mL^{-1}
Enbrel	Etanercept[a]	n/a	Soluble
Remicade	Infliximab[b]	n/a	Soluble
Humira	Adalimumab[c]	n/a	Soluble
Zyprexa	Olanzapine	7.4	Practically insoluble

[a] 51 KDa protein.
[b] 144 KDa antibody.
[c] 148 KDa antibody

equilibrium of Equation (2.1) virtually to the right-hand side, but weak acids will only partially ionise. The equilibrium constant (or ionisation constant in this instance) can be represented by

$$K_a = \frac{[H_3O^+][A^-]}{[HA]} \tag{2.2}$$

Note that water is considered to be present in excess and so does not appear in Equation (2.2). Taking logarithms of both sides yields

$$\log K_a = \log [H_3O^+] + \log [A^-] - \log [HA] \tag{2.3}$$

Reversing the signs gives

$$-\log K_a = -\log [H_3O^+] - \log [A^-] + \log [HA] \tag{2.4}$$

Using the notation p to represent negative logarithm (and noting that $-\log [H_3O^+]$ is represented by pH) gives

$$pK_a = pH + \log [HA] - \log [A^-] \tag{2.5}$$

and so

$$pK_a = pH + \log \frac{[HA]}{[A^-]} \tag{2.6}$$

or

$$pH = pK_a + \log \frac{[A^-]}{[HA]} \qquad (2.7)$$

Equations (2.6) and (2.7) are known as the Henderson–Hasselbalch equations and allow calculation of the extent of ionisation of an acidic drug substance as a function of pH if the pK_a is known. When the pH is significantly below the pK_a (by at least 3 pH units) a weakly acidic drug substance will be completely un-ionised and when the pH is significantly above the pK_a (by at least 3 pH units) a weakly acidic drug substance will be fully ionised (Figure 2.1).

The degree of ionisation will affect the measured solubility because ionised drug substances are generally more soluble in water. Since $[A^-]$ represents the saturated concentration of the ionised drug substance (S_i) and $[HA]$ represents the saturated concentration of the un-ionised drug substance (i.e. the intrinsic solubility, S_o), then

$$pK_a = pH + \log \frac{S_o}{S_i} \qquad (2.8)$$

At any given pH the observed total solubility (S_t) must be the sum of the un-ionised and ionised fractions:

$$S_t = S_o + S_i \qquad (2.9)$$

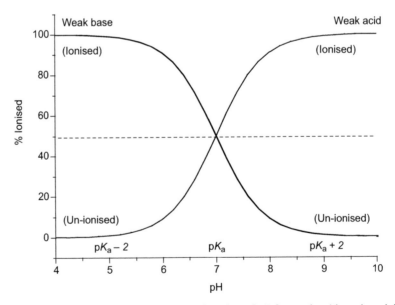

Figure 2.1   Change in percent ionisation as a function of pH for weak acids and weak bases.

Rearranging gives

$$S_i = S_t - S_o \tag{2.10}$$

Substituting in Equation (2.8) gives

$$pK_a = pH + \log \frac{S_o}{S_t - S_o} \tag{2.11}$$

or, in antilog form,

$$S_t = S_o \left[1 + \text{antilog}(pH - pK_a)\right] \tag{2.12}$$

Equation (2.12) allows calculation of the total solubility of an acidic drug substance as a function of pH. Total solubility will be equal to the intrinsic solubility at pH values at least 3 below pK_a and will increase significantly at pH values above pK_a (Figure 2.2). In theory, Equation (2.12) predicts an infinite increase in solubility when $pH \gg pK_a$. In practice this is not attained, real systems deviating from ideal behaviour because of solute–solvent interactions and because salt forms will also have a limiting solubility, but Equation (2.12) is nevertheless a useful approximation over narrow pH ranges.

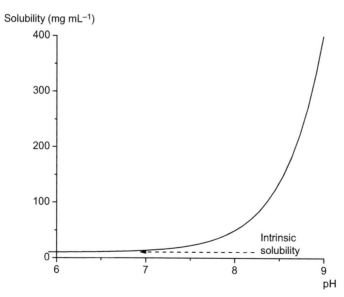

Figure 2.2 Increase in solubility with increasing pH for an acidic drug (pK_a 7.4, S_o 10 mg mL^{-1}).

Example 2.1   The pK_a of the weakly acidic drug substance sulphapyridine is 8.0. What is the ratio of un-ionised to ionised sulphapyridine in blood plasma (pH 7.4)?

From Equation (2.8),

$$\log \frac{S_o}{S_i} = pK_a - pH = 8.0 - 7.4 = 0.6$$

So the ratio of un-ionised to ionised drug must be

$$S_o : S_i = \text{antilog } 0.6 = 3.98 : 1$$

A similar derivation can be made for weak bases. Starting with the notation B to represent the base,

$$B + H_3O^+ \rightleftharpoons BH^+ + H_2O \qquad (2.13)$$

Traditionally a base ionisation constant would be written (K_b) but this has been superseded (somewhat confusingly[1]) by the universal use of acid ionisation constants (i.e. considering the process in Equation (2.13) from right to left) and so

$$K_a = \frac{[H_3O^+][B]}{[BH^+]} \qquad (2.14)$$

If the same logic is applied as above then it is possible to derive

$$pK_a = pH + \log \frac{[BH^+]}{[B]} \qquad (2.15)$$

or

$$pH = pK_a + \log \frac{[B]}{[BH^+]} \qquad (2.16)$$

As in the case of weak acids, the total solubility will be the sum of the solubilities of the ionised and un-ionised moieties and so again following the logic described earlier, the following equations can be described as

$$pK_a = pH + \log \frac{S_t - S_o}{S_o} \qquad (2.17)$$

$$S_t = S_o \left[1 + \text{antilog}(pK_a - pH)\right] \qquad (2.18)$$

---

[1] Confusing because it is not possible to know whether the compound is a weak acid or base without structure or solubility information. Defining a K_b instantly identifies the compound as a weak base.

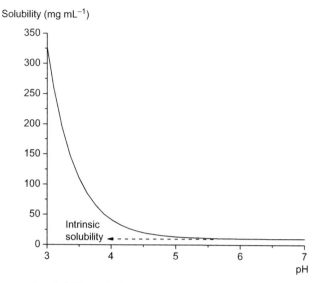

Figure 2.3 Increase in solubility with decreasing pH for a basic drug (pK_a 4.5, S_o 10 mg mL^{-1}).

Equation (2.18) implies that total solubility will be equal to the intrinsic solubility at pH values at least 3 above pK_a and will increase significantly at pH values below pK_a (Figure 2.3). Note again that the solubility profile predicted from Equation (2.18) assumes ideal behavior and that in practice measured solubility often deviates from ideality. The data in Figure 2.4 for metronidazole (pK_a 2.6) illustrate this point.

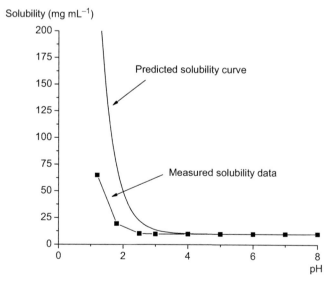

Figure 2.4 Solubility data for metronidazole, showing the difference between experimentally measured solubility data and the curve predicted using Equation (2.18) (solubility data from Wu and Fassihi (2005)).

Study question 2.1   Derive Equations (2.17) and (2.18).

Study question 2.2   Look at the solubility data for Lipitor (atorvastatin calcium) and Seroquel (quetiapine fumarate) in Table 2.1. Can you tell which drug is a weak acid and which is a weak base?

Summary box 2.1

• If a drug substance has an ionisable group, then solubility will vary with pH.

• The ionisation constant is known as the pK_a.

• Acidic drugs will be more soluble when $pH \gg pK_a$.

• Basic drugs will be more soluble when $pH \ll pK_a$.

• When $pH = pK_a$ the drug is 50% ionised.

• The Henderson–Hasselbalch equations allow prediction of solubility as a function of pH.

2.2.1   Percent ionisation

Rearrangement of the Henderson–Hasselbalch equations allows calculation of the percentage of drug substance ionised as a function of pH. Starting with the premise that

$$\text{Percent ionised} = \frac{S_i}{S_i + S_t} \times 100 \tag{2.19}$$

Rearranging the Henderson–Hasselbalch equation for acid species (Equation (2.8)) yields

$$pK_a - pH = \log \frac{S_o}{S_i} \tag{2.20}$$

Taking antilogs gives

$$\text{antilog}(pK_a - pH) = \frac{S_o}{S_i} \tag{2.21}$$

or

$$S_i \, \text{antilog}(pK_a - pH) = S_o \tag{2.22}$$

Substituting Equation (2.22) into Equation (2.19) yields

$$\text{Percent ionised} = \frac{100 \, S_i}{S_i + S_i \, \text{antilog}(pK_a - pH)} \qquad (2.23)$$

or, dividing through by S_i,

$$\text{Percent ionised} = \frac{100}{1 + \text{antilog}(pK_a - pH)} \qquad (2.24)$$

Equation (2.24) (and the equivalent form for basic drug substances – see Study question 2.1) allows calculation of the percent ionised as a function of pH with respect to pK_a. Some values are given in Table 2.2 for reference.

Table 2.2 Percent ionised for weak acids and bases as a function of pH.

	Percent ionised	
pK_a – pH	Weak acid	Weak base
−4	99.99	0.01
−3	99.94	0.06
−2	99.01	0.99
−1	90.91	9.09
−0.9	88.81	11.19
−0.8	86.30	13.70
−0.7	83.37	16.63
−0.6	79.93	20.07
−0.5	75.97	24.03
−0.4	71.53	28.47
−0.3	66.61	33.39
−0.2	61.32	38.68
−0.1	55.73	44.27
0	50	50
0.1	44.27	55.73
0.2	38.68	61.32
0.3	33.39	66.61
0.4	28.47	71.53
0.5	24.03	75.97
0.6	20.07	79.93
0.7	16.63	83.37
0.8	13.70	86.30
0.9	11.19	88.81
1	9.09	90.91
2	0.99	99.01
3	0.06	99.94
4	0.01	99.99

Study question 2.3   Derive an equation that allows determination of per-cent ionisation as a function of pH for basic drug substances.

2.3   Buffers

Clearly, when an acid is added to an aqueous solution the pH will drop and when a base is added to an aqueous solution the pH will rise (this phenomenon is used as the basis of a therapeutic treatment – antacids). Sometimes, it is preferable to have a solution that resists changes in pH upon dissolution of a solute (because solubility, percent ionisation and some reaction rates are pH-dependent). A solution designed to resist changes in pH upon addition of an acid or base is known as a *buffer*. A buffer system usually comprises a weak acid or base and its corresponding *salt*. An example would be a combination of acetic acid and sodium acetate:

$$CH_3COOH \rightleftharpoons CH_3COO^- + H^+$$

$$CH_3COONa \rightleftharpoons CH_3COO^- + Na^+$$

Acetic acid is a weak acid and therefore its position of equilibrium lies to the left. Sodium acetate is a salt and will dissociate fully, so its equilibrium lies to the right. If H^+ ions are added they will react with CH_3COO^- ions to give acetic acid. This is a weak acid and so only ionises slightly. The pH thus remains practically constant. Similarly, if OH^- ions are added they react with acetic acid to give water and acetate ions. Again, the pH only changes slightly. Buffers usually work best when the pH being maintained is equal or near to the pK_a of the weak acid or base used. Buffer capacity (the number of moles of acid or base that must be added to a buffer system to produce a decrease or increase of 1 pH unit) can be determined from the Henderson–Hasselbalch equation:

$$pH = pK_a + \log \frac{[\text{salt}]}{[\text{acid}]} \qquad (2.25)$$

2.4   Determination of pK_a

There are numerous methods to measure pK_a. Prankerd (2007) provides an excellent introduction to the methods available as well as comprehensive tables of pK_a data for existing drugs and notes the most accurate methods are those based on conductance (reliable to ± 0.0001 pK unit or better) or electrochemical cells without liquid junction potentials (reliable to ± 0.001 pK unit or better). Pharmaceutical pK_a values are usually determined with methods based on a relationship between solution pH and a physicochemical

Table 2.3 IUPAC descriptions for error in pK_a values.

Description	Error in pK_a	Uncertainty in pK_a
Very reliable	$< \pm 0.005$	$<1\%$
Reliable	± 0.005 to ± 0.02	$\sim 1\%$
Approximate	± 0.02 to ± 0.04	$\sim 5\%$
Uncertain	$> \pm 0.04$	$>10\%$

property, such as added titrant concentration, spectrophotometric absorbance, optical rotation or fluorescence intensity. Prankerd (2007) notes that using any approach based on measurement of pH immediately limits the accuracy and precision of the result to no better than ± 0.02 pK unit.

IUPAC has published descriptive terms for the error of pK_a measurements (Table 2.3). Although the errors appear small, it must be remembered that pK_a expresses the ionisation constant in logarithmic terms (the corresponding uncertainties in K_a are also shown in Table 2.3). Determination of pK_a values with a method based on measurement of pH can therefore only result in a value that can be considered approximate at best and most likely uncertain. As noted earlier, during preformulation screening it is acceptable to determine approximate values since the data are used primarily to identify those drug substances with the best potential for development. Hence, initial pK_a values should be regarded as uncertain.

2.4.1 Determination of pK_a by potentiometric titration

Modern automated instrumentation is available that can determine pK_a values, by potentiometric titration, with very small (<10 mg) amounts of drug substance. The drug substance is dissolved in water, forming either a weakly acidic or weakly basic solution. Acid or base (as appropriate) solution is titrated and the solution pH recorded. A plot of volume of titrant solution added versus pH allows graphical determination of the pK_a, since when pH $=$ pK_a the compound is 50% ionised (Figure 2.5). The method has the significant advantage of not requiring an assay for the analyte.

2.4.2 Determination of pK_a in nonaqueous solvents

One drawback of these methods is that if a drug substance has poor aqueous solubility then differences in pH because of ionisation can be difficult to measure. Prankerd (2007) suggests solubility of 0.001 M (0.2 mg mL^{-1} for a compound of molecular weight 200 Da) as a practicable lower limit. In cases where solubility is too low for pK_a to be determined potentiometrically, an organic co-solvent can be used to increase solubility (Takács-Novák, Box and

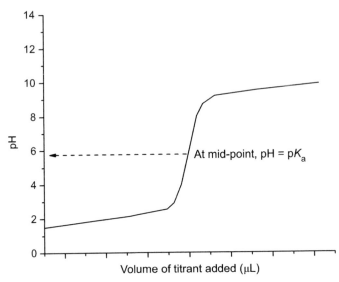

Figure 2.5 A plot of volume of titrant added versus solution pH for a potentiometric pH titration.

Avdeef, 1997). Typically methanol is used, since it is miscible with water, shows a solvation effect close to water and solubility often approaches ideality (see Chapter 3). Addition of methanol (or any other co-solvent) changes the polarity of the solvent. This in turn will affect dissociation and so pK_a values determined in nonaqueous solvents are termed *apparent* (pK'_a). Generally dissociation of weak acids is more affected by solvent polarity that dissociation of weak bases.

Determining pK'_a values in a series of solvents of increasing co-solvent concentration allows extrapolation of the pK_a value in pure water. This approach increases experimental time, but equally increases confidence in the pK_a value obtained. Several methods have been reported for achieving extrapolation, the most common being that developed independently by Yasuda (1959) and Shedlovsky (1962):

$$pK'_a + \log[H_2O] = \frac{A}{\varepsilon} + B \qquad (2.26)$$

where ε is the dielectric constant of the co-solvent mixture and A and B are constants. Accordingly, a plot of $pK'_a + \log[H_2O]$ versus $1/\varepsilon$ should be a straight line of slope A and intercept B. The pK_a in water is calculated when $[H_2O] = 55.5$ molal and $\varepsilon = 78.5$ (Table 2.4). Figure 2.6 shows Yasuda–Shedlovsky plots for a number of drug substances. Table 2.4 gives the extrapolated and experimentally determined (in water) pK_a values.

Table 2.4 Measured (in water) and extrapolated (from the Yasuda–Shedlovsky plots) pK_a values for three drugs (data from Takács-Novák et al. (1997)).

Drug	pK_a (water)	pK_a (extrapolated)	ΔpK_a	Description
Salicylic acid	2.75	2.73	0.02	Approximate
Phenobarbital	7.41	7.41	0.0	Very reliable
Paracetamol	9.63	9.67	0.04	Uncertain
	9.60	9.60	0.0	Very reliable

2.4.3 Other factors affecting measurement of pK_a

The ionic strength of the solution in which measurements are made will generally affect the activity of all dissolved ionic species and so will influence pK_a. It is possible to use the Debye–Hückel equations to correct for ionic strength effects. Since biological fluids may have high ionic strengths, pK_a values may be better determined in simulated media.

Dissolved carbon dioxide, by virtue of forming carbonic acid upon dissolution, can also affect pK_a if a potentiometric titration is used. The first pK_a of carbonic acid is 6.2, so there will be no effect for acids or conjugate acids of pK_a lower than 6.2, but a significant effect may occur where the pK_a of the analyte is greater than 6.2 because the method requires input of the volumes of acid or base required to cause neutralisation.

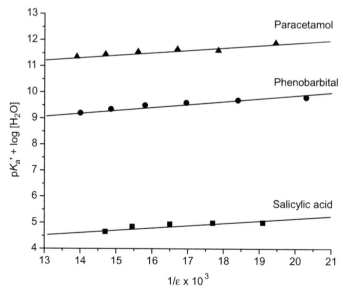

Figure 2.6 Yasuda–Shedlovsky plots for three drugs (redrawn from Takács-Novák et al., Copyright (1997), with permission from Elsevier).

Photolysis, oxidation and hydrolysis may also affect the measured pK_a, depending upon the properties of any degradation products formed. The effects of photolysis and oxidation can be mitigated through careful experimental design; where hydrolysis is suspected, the experimental time must be kept to a minimum.

Summary box 2.2

- Pharmaceutical pK_a values are usually determined with a method based on measurement of pH and so must be considered uncertain.

- Commercial instruments usually determine pK_a potentiometrically.

- Apparent pK_a values can be determined in nonaqueous solvents.

- For poorly soluble compounds, pK_a values can be determined by addition of various amounts of a co-solvent and using a Yasuda–Shedlovsky extrapolation.

2.5 Summary

Where a drug has an ionisable functional group solubility will change with pH. Acidic drugs will be freely soluble in high pH media and basic drugs will be freely soluble in low pH media. Ionisation is characterised by the pK_a value. When $pH = pK_a$ the molecule will be half-ionised. Knowledge of pK_a allows prediction of solubility with pH and also aids salt selection, should this be deemed necessary (Chapter 6). During preformulation screening, determination of pK_a potentiometrically is most convenient, but it must be recognised that the value so produced should be treated as uncertain.

References

Manallack, D.T. (2007) The pK_a distribution of drugs: application to drug discovery. *Perspect. Med. Chem.*, **1**, 25–38.

Prankerd, R.J. (2007) Critical compilation of pK_a values for pharmaceutical substances. In *Profiles of Drug Substances, Excipients and Related Methodology* (ed. H.B. Brittain), vol. 33, Academic Press, San Diego, pp. 1–726. ISBN 9780-1226-0833-9.

Shedlovsky, T. (1962) The behaviour of carboxylic acids in mixed solvents. In *Electrolytes* (ed. B. Pesce), Pergamon Press, pp. 146–151. ISBN 0-0801-3778-4.

Takács-Novák, K., Box, K.J. and Avdeef, A. (1997) Potentiometric pK_a determination of water-insoluble compounds: validation study in methanol/water mixtures. *Int. J. Pharm.*, **151**, 235–248.

Wells, J.I. (1988) Pharmaceutical preformulation. In *The Physicochemical Properties of Drug Substances*, John Wiley & Sons, Ltd, Chichester. ISBN 0-470-21114-8.

Wu, Y. and Fassihi, R. (2005) Stability of metronidazole, tetracycline HCl and famotidine alone and in combination. *Int. J. Pharm.*, **290**, 1–13.

Yasuda, M. (1959) Dissociation constants of some carboxylic acids in mixed aqueous solvents. *Bull. Chem. Soc. Jpn*, **32**, 429–432.

Answers to study questions

2.1 In this instance $[BH^+]$ represents the saturated concentration of ionised drug substance (S_i) and $[B]$ represents the saturated concentration of un-ionised drug substance (i.e. the intrinsic solubility, S_o). Therefore, from Equation (2.15),

$$pK_a = pH + \log\frac{S_i}{S_o}$$

At any given pH the observed total solubility (S_t) must be the sum of the un-ionised and ionised fractions:

$$S_t = S_o + S_i$$

Rearranging,

$$S_i = S_t + S_o$$

Substituting gives

$$pK_a = pH + \log\frac{S_t - S_o}{S_o}$$

or, in antilog form,

$$S_t = S_o\left[1 + antilog(pK_a - pH)\right]$$

2.2 Atorvastatin has greater solubility at higher pH and so is weakly acidic, while quetiapine has greater solubility at lower pH and so is weakly basic. Interestingly, as both are marketed as salt forms, atorvastatin is present as a conjugate base and quetiapine is present as a conjugate acid. See Chapter 6 for further discussion of this principle.

2.3 Starting with Equation (2.16),

$$pH = pK_a + \log \frac{[B]}{[BH^+]}$$

$$pH = pK_a + \log \frac{S_o}{S_i}$$

$$pH - pK_a = \log \frac{S_o}{S_i}$$

Taking antilogs,

$$antilog(pH - pK_a) = \frac{S_o}{S_i}$$

or

$$S_i \; antilog(pH - pK_a) = S_o$$

Substituting into Equation (2.19) yields

$$Percent \; ionised = \frac{100 \; S_i}{S_i + S_i \; antilog(pH - pK_a)}$$

or, dividing through by S_i,

$$Percent \; ionised = \frac{100}{1 + antilog(pH - pK_a)}$$

Additional self-study questions and answers

2.1 The pK_a of the weakly acidic drug sulphapyridine is 8.0. What is the ratio of un-ionised to ionised sulphapyridine in gastric contents (pH 2.0)? From Equation (2.8),

$$\log \frac{S_o}{S_i} = pK_a - pH = 2.0 - 7.4 = -5.4$$

Therefore the ratio of un-ionised to ionised drug must be

$$S_o : S_i = antilog - 5.4 = 3.98 \times 10^{-6} : 1$$

2.2 The pK_a of the weakly acidic drug sulphapyridine is 8.0. At what pH does the ratio of un-ionised to ionised sulphapyridine become 1:1?

$$\log \frac{S_o}{S_i} = \log 1 = 0$$

$$0 = pK_a - pH \text{ so } pH = pK_a - 0 = 8.0 - 0 = 8.0$$

When the pH equals the pK_a the drug is 50% ionised.

3
Partition Affinity

3.1 Introduction

Once in solution a drug substance must be absorbed into the bloodstream. This involves transport across biological membranes. Membranes are typically lipophilic in nature, so a drug that is highly soluble in water will not be soluble in such a hydrophobic environment. Equally, a drug with low aqueous solubility will probably have high solubility in a lipophilic membrane. This presents a considerable challenge to successful drug delivery. While there may be active transport mechanisms to uptake the drug substance, general absorption can be approximated by understanding the relative affinity of the compound between water and organic phases. This behaviour, partitioning, is another defining characteristic of a drug substance that, together with solubility, indicates likely developability.

3.2 Partitioning

When a solute is added to a mixture of two (immiscible) solvents it will usually dissolve in both to some extent and a position of equilibrium will be established between the concentrations (C) in the two solvents. In other words, the ratio of the concentrations will be constant and is given by

$$P_{1,2} = \frac{C_1}{C_2} \tag{3.1}$$

where P is the *partition coefficient* and the subscripts indicate the solvent phase. Note that it would be equally possible to define

$$P_{2,1} = \frac{C_2}{C_1} \tag{3.2}$$

Essentials of Pharmaceutical Preformulation, First Edition. Simon Gaisford and Mark Saunders.
© 2013 John Wiley & Sons, Ltd. Published 2013 by John Wiley & Sons, Ltd.

In a physiological environment drug substances partition from an aqueous phase to numerous and heterogeneous lipophilic phases (typically cell membranes). It would be complex to develop an analytical method that allowed measurement of partitioning between such complex phases and so a simple solvent model is usually used instead (typically n-octanol, which is taken to mimic the short-chain hydrocarbons that make up many biological lipid bilayers). Thus, a partition coefficient can be written to represent the distribution of a solute between water (w) and n-octanol (o):

$$P_{o,w} = \frac{C_o}{C_w} \qquad (3.3)$$

Again, the following could be defined:

$$P_{w,o} = \frac{C_w}{C_o} \qquad (3.4)$$

However, by convention $P_{o,w}$ is the standard term (easily remembered as organic phases usually sit above water, since they are less dense).

Study question 3.1   What range will the values of $P_{o,w}$ have for lipophilic and hydrophilic drug substances?

For simplicity it will be assumed that only un-ionised species can partition into the organic phase[1] so the partition coefficient as defined above applies only if (a) the drug substance cannot ionise or (b) the pH of the aqueous phase is such that the drug substance is completely un-ionised. If the drug substance has partially ionised in the aqueous phase and partitioning is measured experimentally then the parameter measured is the distribution coefficient, D:

$$D_{o,w} = \frac{C_o}{C_{w,ionised} + C_{w,un\text{-}ionised}} \qquad (3.5)$$

The partition coefficient and the distribution coefficient are related by the fraction of solute un-ionised ($f_{un\text{-}ionised}$):

$$D_{o,w} = f_{un\text{-}ionised} P_{o,w} \qquad (3.6)$$

---

[1] In general, ionised species are too polar to dissolve in organic phases, although some protonated species (amines for instance) partition significantly into chloroform, which is why ether is used preferentially for quantitative extraction assays.

Summary box 3.1

- Partition coefficients are determined between water and an organic (usually *n*-octanol) phase.

- Partition coefficients can be defined as o/w or w/o.

- By convention, o/w values are used.

- For lipophilic drugs, $P > 1$ or log P is positive.

- For hydrophilic drugs, $P < 1$ or log P is negative.

- Ionised drug has low affinity for organic phases.

- Experimental measurement of partitioning for a drug that ionises results in a distribution coefficient.

Note that partition coefficients may be defined between any organic phase and water. *n*-Octanol is a common choice but it is by no means either the best choice or the only choice, especially if partition coefficients are estimated using chromatographic methods. The Collander equation relates the partition coefficients for water with a range of organic solvents:

$$\ln P_\mathrm{I} = a + b \ln P_\mathrm{II} \tag{3.7}$$

where the subscripts I and II refer to water and the organic solvent respectively and a and b are constants. Beezer *et al.* (1987) show that while the constant a has no fundamental significance, the value of b is reflective of differences between organic phases and is useful in scaling solvent behaviour to biological membrane properties.

3.2.1   Effect of partitioning

Leo, Hansch and Elkins (1971) review the use of partition coefficients in pharmaceutical development. The main effect, from the perspective of a drug substance, is on absorption *in vivo*. Crossing biological membranes (assuming no active transport mechanisms are available) essentially involves a series of partitioning steps between organic and aqueous phases and so a balance of lipophilicity and hydrophilicity is required to prevent accumulation of the drug substance in one particular phase. The optimum balance depends upon the site of absorption within the body. Buccal and GI tract absorption require a relatively high hydrophilicity, while a low value is beneficial in aiding uptake across the blood–brain barrier.

Partitioning will also affect solubility when two phases are in contact with each other and, since ionic species have little affinity for organic phases, the bioavailability of salt forms may be limited. From an analytical perspective partitioning also underpins solvent extraction techniques.

3.2.2 Determination of log *P*

Values for log *P* can be determined experimentally or can be estimated using group additivity functions. For the latter approach there are numerous computer models and simulation methods available and selection will reduce to personal choice and familiarity, and so these models will not be considered here. This text will focus on experimental determination. It is clear, however, that there is much value to be gained by comparing calculated and experimentally determined log *P* values; nor can the value of the calculational approach be underestimated when selecting a lead candidate from a compound library when it would simply not be either possible or practicable to measure the partitioning behaviour of the many thousands of compounds available.

3.2.2.1 Shake-flask method Assuming a UV–visible assay is available, then the shake-flask method is a quick, simple and near universally applicable way of determining the partition coefficient. Prior to measurement, the solvents to be used (normally water, or buffer, and *n*-octanol) should be mixed with each other and allowed to reach a position of equilibrium. This is because each solvent has a significant solubility in the other (*n*-octanol in water: 4.5×10^{-3} M; water in *n*-octanol: 2.6 M, Weber, Chin and Rice, 1986). There can be a significant effect of saturating water with *n*-octanol on the solubilities of compounds. Chiou, Schmedding and Manes (1982) report an increase in solubility of DDT and hexachlorobenzene of 160 and 80% respectively in water saturated with *n*-octanol compared with the solubilities in pure water.

Following preparation of the solvent phases, the drug substance is dissolved in the aqueous phase to a known concentration. This could be the solubility (indeed, the solution used to determine solubility can subsequently be used in the partitioning experiment) because the higher the concentration the more reliable the assay. Equal volumes of aqueous drug solution and *n*-octanol are then mixed in a separating funnel. The mixture should be shaken vigorously for a period of time (usually 30 min, to maximise the surface area of the two solvents in contact with each other) while the drug substance

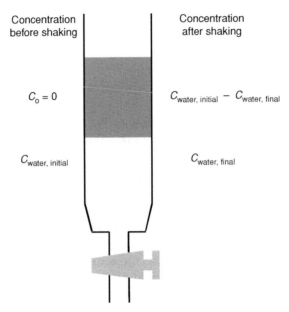

Figure 3.1   The shake flask method for determination of partition coefficient.

partitions. The phases are allowed to separate (5 min) and then the concentration of drug substance remaining in the aqueous phase is determined with the UV–visible assay already developed (Figure 3.1). By difference, the concentration of drug substance in the n-octanol phase is known:

$$C_{n\text{-octanol}} = C_{\text{water,initial}} - C_{\text{water,final}} \qquad (3.8)$$

When the partition coefficient heavily favours distribution to the n-octanol phase then a smaller volume of n-octanol can be added, since this will increase the concentration in the aqueous phase at equilibrium and reduce the error in the analytical determination of the concentration. However, the calculation for partition coefficient needs to be corrected to account for the different volumes. Assuming a 1:9 n-octanol:water ratio, then Equation (3.1) becomes

$$P_{o,w} = \frac{10C_o}{C_w} \qquad (3.9)$$

Study question 3.2   What are the drawbacks of the shake-flask method?

n-Octanol may also not be the best organic phase. Hexane or heptane can be used as an alternative, although they will give different partition coefficient values from *n*-octanol and are also considered to be less representative of biological membranes because they cannot form any hydrogen bonds with the solute. Where the aim of the experiment is to differentiate partitioning between members of a homologous series the organic phase can be varied in order to maximise discrimination. *n*-Butanol tends to result in similar partition values for a homologous series of solutes, while heptane tends to exaggerate differences in solute lipophilicity. Solvents that are more polar than *n*-octanol are termed *hypodiscriminating* and those that are less polar than *n*-octanol are termed *hyperdiscriminating*. Hyperdiscriminating solvents reflect more closely transport across the blood–brain barrier while hypodiscriminating solvents give values consistent with buccal absorption. *n*-Octanol tends to reflect absorption from the GI tract, which is why it is the default option. The discriminating powers of a range of common solvents, relative to *n*-octanol, are shown in Figure 3.2.

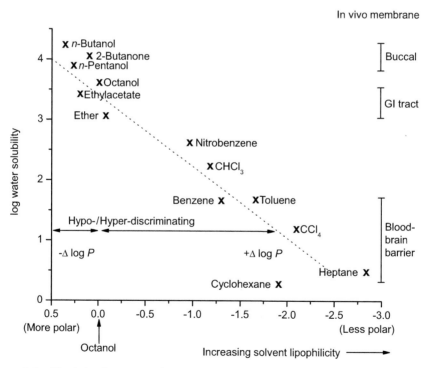

Figure 3.2 Discriminating power of various partitioning solvents (redrawn from Wells (1988), with permission from John Wiley & Sons, Inc.).

Summary box 3.2

- n-Octanol is usually selected as the organic phase as it is taken to be representative of absorption across the GI tract.

- Solvents that are more polar than n-octanol are termed hypodiscriminating.

- Solvents that are less polar than n-octanol are termed hyperdiscriminating.

- Hyperdiscriminating solvents reflect more closely transport across the blood–brain barrier.

- Hypodiscriminating solvents give values consistent with buccal absorption.

- Partition coefficients between water and organic phases can be related through the Collander equation.

3.2.2.2 Chromatographic methods Separation of analytes by liquid chromatographic methods relies on an interaction between the analytes (dissolved in a mobile phase) and a stationary (solid) phase. In normal-phase chromatography the stationary phase is polar and the mobile phase is nonpolar and in reverse-phase (RP) chromatography the stationary phase is nonpolar and the mobile phase is polar. It follows that liquid chromatography can be used with single analytes to measure partitioning behaviour, since the extent of interaction must depend upon the relative lipophilicity or hydrophilicity of the analyte. Typically, reverse-phase chromatography is used for partitioning experiments (Lambert, 1993).

Reverse-phase TLC allows measurement of partition coefficients (Biagi *et al.*, 1964, 1994a, 1994b, 1994c) by comparing progression of a solute relative to progression of the solvent front (the ratio of the two being the resolution factor, R_f). The resolution factor achieved for each drug is converted to a TLC retention factor (R_m), which is proportional to log P:

$$R_m = \log\left(\frac{1}{R_f} - 1\right) \qquad (3.10)$$

The stationary phase can be n-octanol but is more commonly silica impregnated with silicone oil. The mobile phase can in principle be water (or buffer) but unless the solute is reasonably hydrophilic good resolution tends not to be achieved with water alone and reasonably lipophilic compounds tend not to move from the starting line at all (i.e. $R_f = 0$). Co-solvents (typically acetone, acetonitrile or methanol) are added to the mobile phase to increase the migration of highly lipophilic compounds. The nearer the compound migrates to the solvent front the higher the resolution factor (the maximum value attainable being 1). The value of R_f can thus vary between 0 and 1, corresponding to R_m values from $+\infty$ to $-\infty$ respectively, although in practice the measurable range is usually 0.03 to 0.97, corresponding to R_m values of 1.5 to -1.5 respectively (Biagi et al., 1994a).

Addition of a co-solvent can be used to modulate the value of R_f obtained and the relationship is usually linear. This being so, it is possible to extrapolate to zero co-solvent and so calculate R_m in water. Biagi et al. (1994a) demonstrate that the agreement between extrapolated and measured R_f values is excellent for hydrophilic compounds (that migrate to some extent in water and thus have a measurable value in water alone).

There are many examples in the literature where reverse-phase TLC has been applied to pharmaceuticals. Bird and Marshall (1971) show that the R_m values for a series of eight penicillins are in near-perfect relation to the log P values of the penicillin-free acids determined in n-octanol and water. They also show that the resolving factor is pH-dependent, achieving satisfactory separation when the pH of the mobile phase is pH 3 or 4, but that the resolving power is lost at pH 5 and higher. Biagi et al. (1969) discuss the use of reverse-phase TLC for determining the partitioning behavior of cephalosporins, while Jack et al. (1989) used the technique to study partitioning of three classes of drug: β-adrenoceptor antagonists, nonsteroidal anti-inflammatory agents and dihydropyridine calcium antagonists. A more recent study has reviewed the use of reverse-phase TLC for lipophilicity determination using 35 drugs (Komsta et al., 2010).

Reverse-phase TLC offers the advantage of using small solvent and solute volumes as well as the ability to measure several solutes simultaneously, but requires preparation of the plate as well as developing to reveal the position of the solute. Reverse-phase HPLC is an alternative, and widely used, technique for measurement of partition coefficients. The stationary phase comprises a nonpolar compound (typically a C_{18} hydrocarbon) chemically bound to an inert, solid support medium (such as silica). The mobile phase is usually an admixture of water with acetonitrile or methanol and can be isocratic or nonisocratic (see below). It is possible to use water saturated with n-octanol as the mobile phase and a stationary phase covered in n-octanol, but the eluting power is not strong, for the same reason noted above for TLC, and so to measure an acceptable range of partition coefficients it is necessary to change the volume ratio of mobile to stationary phase.

Conceptually, it is not clear whether the interaction between the solute and the stationary phase constitutes surface adsorption or true phase partitioning. The longer carbon-chain hydrocarbons (relative to *n*-octanol) have been found to provide a better correlation to log *P* values, presumably because their greater reach from the solid surface of the support matrix means they behave as a more liquid-like phase. Nonetheless, the number of degrees of freedom of movement of the alkyl groups must be reduced when bound to a solid substrate, so true partitioning is unlikely (Weber, Chin and Rice, 1986).

The interaction between solute and stationary phase is characterised by the retention factor (k). The greater the retention factor, the longer the solute takes to elute from the column (the retention time, t_r):

$$k = \frac{t_r - t_0}{t_0} = \frac{n_s}{n_m} \tag{3.11}$$

where t_0 is the dead time (the time taken for the solvent to elute from the column). Empirically, there is often strong correlation between the retention factor and log *P* values determined from octanol–water. Figure 3.3 shows the correlation for three series of barbituric acids.

The retention factor is also given by the ratio of the number of molecules of solute in the stationary phase (n_s) to the number of molecules of solute in the mobile phase (n_m). The greater the affinity of the solute for the stationary phase, the greater the volume of mobile phase required for elution, and so the

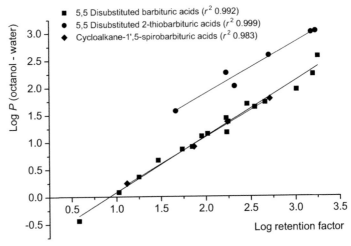

Figure 3.3   Log *P* versus log retention factor correlation (linear regression coefficients are shown as r^2 values) for three series of barbituric acids (data courtesy of Richard Prankerd).

retention factor is found to be related to the partition coefficient by (Valkó, 2004)

$$\log k = \log P + \left(\frac{V_s}{V_m}\right) \qquad (3.12)$$

So long as the volumes of the mobile and stationary phases are known, Equations (3.11) and (3.12) show that partition coefficients can be determined from HPLC measurements from retention time data alone (i.e. there is no need to determine concentrations). Additional benefits include the fact that the concentration of solute injected on to the column does not affect the retention time and any impurities present are naturally separated during the course of the experiment. In addition, if the compound has poor aqueous solubility it can be dissolved in another solvent prior to being injected to the column – the solvent will be separated during the experiment and the compound eluted with an aqueous mobile phase.

As noted above, a co-solvent is often added to the mobile phase to encourage elution of lipophilic compounds. If a single ratio of co-solvent to water is used the method is known as *isocratic* and if the ratio is changed during the experiment the method is known as *gradient*. Isocratic methods are easier to run but require knowledge of the likely lipophilicity of the compound prior to experimentation. Since this information is unlikely to be known (it is, in fact, the reason for performing the experiment in the first place) a range of co-solvent concentrations must be tested, but in any event gradient methods are usually performed first in any HPLC method development programme. In order to compare results for a series of compounds, the same co-solvent concentration must be used or the data must be extrapolated to the same (or zero) concentration. Usually a linear relationship is found between co-solvent concentration and k:

$$\log k = aC + \log k_w \qquad (3.13)$$

where C is the concentration of co-solvent, a is a constant of proportionality and k_w is the value of k in pure water. In principle a plot of $\log k$ versus C should result in a straight line from which k_w can be determined. In practice such plots are usually acceptably linear only within the range $-0.5 < \log k < 1.5$ (Valkó, 2004). Also, since the measured retention times will vary depending upon the column, it is necessary to validate the data by measuring a set of reference materials of known $\log P$. However, the model does not generally extrapolate over a range of structurally diverse compounds, which is manifest in different values of a in Equation (3.13) (in other words, the point of interception of the lines generated by Equation

(3.13) is often not zero co-solvent). This being so, the value of a can be used as a second variable in relating the HPLC data to log P:

$$\log P = xa + y \log k_w + z \tag{3.14}$$

where x, y and z are regression coefficients (Valkó, 1984). The ratio x/y gives the concentration of co-solvent to which the data should be extrapolated.

The ratio $-\log k_w/a$ has also been shown to correlate better with log P values than k_w alone (Valkó and Slégel, 1993):

$$\log P = d \left(\frac{-\log k_w}{a} \right) + e \tag{3.15}$$

where d and e are regression coefficients. The ratio $-\log k_w/a$ corresponds to the co-solvent concentration that is required to get the solute retention time to be exactly double the dead time (i.e. $\log k = 0$).

In a fast-gradient method the ratio of co-solvent (usually acetonitrile) to water is linearly increased with time, leading to faster elution (typically less than 5 min). If the dead volume and dwell volume of the column are known, then it is possible to estimate the co-solvent concentration as the compound elutes from the column. The retention times during a fast-gradient run should be approximately equal to the ratio $-\log k_w/a$. Valkó, Bevan and Reynolds (1997) demonstrated the validity of this approach for 76 diverse drug molecules.

3.2.3   Effect of salt formation on partitioning

The formation of a salt almost always infers greater hydrophilicity on a base compound (Table 3.1). It can be seen that for a range of weakly basic drug substances, the formation of the corresponding hydrochloride salt reduces the $P_{o/w}$ of the drugs by between 100 and 10 000 (i.e. the lipophilicity falls by

Table 3.1   The effect of salt formation on the partition coefficient of weak bases.

	log P		
Base	Free base	Hydrochloride salt	Δ log P
Chlorpromazine	5.35	1.51	3.84
Promazine	4.49	0.91	3.58
Triflupromazine	5.19	1.78	4.28
Trifluperazine	5.03	1.69	3.34
Diphenhydrazine	3.30	−0.12	3.42
Propranolol	3.18	−0.45	3.63
Phenylpropranolamine	1.83	−1.09	2.92

2 to 4 orders of magnitude on salt formation). The latter clearly explains the increased water solubility observed with hydrochloride salts.

Summary box 3.3

- RP-TLC or RP-HPLC can be used to estimate partition coefficients.

- The formation of a salt almost always infers greater hydrophilicity on a base compound.

3.3 Summary

Partition coefficients provide a quantitative value to correlate with permeability and so are useful in characterising the likely developability of a drug substance. The conventional method of determination is to measure distribution between water and *n*-octanol, although other solvents can be used, especially when trying to understand permeability across the blood–brain barrier. RP-TLC and HPLC can also be used to estimate partition coefficients, since progression of a solute in the mobile phase of either technique effectively occurs via a series of partitioning steps with the stationary phase.

References

Beezer, A.E., Gooch, C.A., Hunter, W.H. and Volpe, P.L. (1987) A thermodynamic analysis of the Collander equation and establishment of a reference solvent for use in drug partitioning studies. *J. Pharm. Pharmacol.*, **39**, 774–779.

Biagi, G.L., Barbaro, A.M., Gamba, M.F. and Guerra, M.C. (1964) Partition data of penicillins determined by means of reversed-phase thin-layer chromatography. *J. Chromatogr. A*, **41**, 371–379.

Biagi, G.L., Barbaro, A.M., Guerra, M.C. and Gamba, M.F. (1969) Partition data of cephalosporins determined by means of reversed-phase thin layer chromatography. *J. Chromatogr. A*, **44**, 195–198.

Biagi, G.L., Barbaro, A.M., Sapone, A. and Recanatini, M. (1994a) Determination of lipophilicity by means of reversed-phase thin-layer chromatography. i. Basic aspects and relationship between slope and intercept of TLC equations. *J. Chromatogr. A*, **662**, 341–361.

Biagi, G.L., Barbaro, A.M., Sapone, A. and Recanatini, M. (1994b) Determination of lipophilicity by means of reversed-phase thin-layer chromatography. ii. Influence of the organic modifier on the slope of the thin-layer chromatographic equation. *J. Chromatogr. A*, **669**, 246–253.

Biagi, G.L., Barbaro, A.M. and Recanatini, M. (1994c) Determination of lipophilicity by means of reversed-phase thin-layer chromatography. iii. Study of the TLC equations for a series of ionizable quinolone derivatives. *J. Chromatogr. A*, **678**, 127–137.

Bird, A.E. and Marshall, C. (1971) Reversed-phase thin layer chromatography and partition coefficients of penicillins. *J. Chromatogr. A*, **63**, 313–319.

Chiou, C.T., Schmedding, D.W. and Manes, M. (1982) Partitioning of organic compounds in octanol–water systems. *Environ. Sci. Technol.*, **16**, 4–10.

Jack, D.B., Hawker, J.L., Rooney, L. *et al.* (1988) Measurement of the distribution coefficients of several classes of drug using reversed-phase thin-layer chromatography. *J. Chromatogr. A*, **452**, 257–264.

Komsta, L., Skibinski, R., Berecka, A. *et al.* (2010) Revisiting thin-layer chromatography as a lipophilicity determination tool – a comparative study on several techniques with a model solute set. *J. Pharm. Biomed. Anal.*, **53**, 911–918.

Lambert, W.J. (1993) Modeling oil-water partitioning and membrane permeation using reversed-phase chromatography. *J. Chromatogr. A*, **656**, 469–484.

Leo, A., Hansch, C. and Elkins, D. (1971) Partition coefficients and their uses. *Chem. Rev.*, **71**, 525–616.

Valkó, K. (1984) General approach for the estimation of octanol/water partition coefficient by reversed-phase high-performance liquid chromatography. *J. Liq. Chromatogr.*, **7**, 1405–1424.

Valkó, K. (2004) Application of high-performance liquid chromatography based measurements of lipophilicity to model biological distribution. *J. Chromatogr. A*, **1037**, 299–310.

Valkó, K. and Slégel, P. (1993) New chromatographic hydrophobicity index (ϕ_0) based on the slope and the intercept of the log K' versus organic phase concentration plot. *J. Chromatogr. A*, **631**, 49–61.

Valkó, K., Bevan, C. and Reynolds, D. (1997) Chromatographic hydrophobicity index by fast-gradient RP-HPLC: a high-throughput alternative to log *P*/log *D*. *Anal. Chem.*, **69**, 2022–2029.

Weber Jr, W.J., Chin, Y.-P. and Rice, C.P. (1986) Determination of partition coefficients and aqueous solubilities by reverse phase chromatography – I. Theory and background. *Water Res.*, **20**, 1433–1442.

Wells, J.I. (1988) *Pharmaceutical Preformulation. The Physicochemical Properties of Drug Substances*, John Wiley & Sons, Ltd, Chichester. ISBN 0-470-21114-8.

Answers to study questions

3.1 When a drug substance is lipophilic (i.e. it has a high affinity for the octanol phase) the value of $P_{o,w}$ will be greater than 1 and when a drug substance is hydrophilic the value of $P_{o,w}$ will be less than 1. Since hydrophilic drug substances will give small $P_{o,w}$ values, log $P_{o,w}$ values are often quoted (abbreviated to log *P*), in which case hydrophilic drug substances will have a negative value and lipophilic drug substances a positive value.

3.2 One is that the volumes of solution are reasonably large and another is that enough time must be allowed to ensure that equilibrium partitioning is attained. One way to make sure equilibrium has been reached is to measure the change in concentration (in either phase but typically the aqueous phase) with time until a constant value is seen.

4
Solubility

4.1 Introduction

In order to be absorbed by the body generally a drug substance must be in solution, yet many drugs are formulated in the solid state (usually for good reasons, such as stability and ease of manufacture, transportation and administration). Understanding the solid-state properties of a drug substance provides the foundation upon which to develop the dosage form but understanding the processes by which the drug transitions from the solid state into solution and the equilibrium concentration (thermodynamic solubility) or maximum concentration (kinetic solubility) attained are critical in predicting, and optimising, drug product performance.

It has been estimated that historically up to 40% of drug candidates have been abandoned because of poor aqueous solubility (Kennedy, 1997) and between 35 and 40% of compounds currently in development have aqueous solubilities below 5 mg mL^{-1} at pH 7 (Stegemann *et al.*, 2007). A popular system for categorising drugs on the basis of physicochemical properties is the biopharmaceutical classification system (BCS) (Amidon *et al.*, 1995). The BCS considers solubility and intestinal permeability (defined as the ratio of drug absorbed through the GI tract following oral administration to drug administered intravenously) and organises drug substances into one of four categories (Table 4.1). As such, it is a useful guide to the likely ease of developability of a drug substance.

Highly soluble drug substances are defined as those where the highest dose strength available is dissolvable in <250 mL of water over a pH range 1–7.5 while highly permeable drugs are defined as those where the extent of absorption in humans is greater than 90%, based on a mass–balance analysis or in comparison to an intravenously administered dose. Solubility improvement is thus one development strategy for enhancing oral bioavailability of

Essentials of Pharmaceutical Preformulation, First Edition. Simon Gaisford and Mark Saunders.
© 2013 John Wiley & Sons, Ltd. Published 2013 by John Wiley & Sons, Ltd.

Table 4.1   BCS categories based on solubility and intestinal permeability.

BCS class	Solubility	Intestinal permeability
1	High	High
2	Low	High
3	High	Low
4	Low	Low

BCS class 2 and 4 drugs. Importantly, the US Food and Drug Administration (FDA) permits a BCS biowaiver (i.e. *in vitro* dissolution data are sufficient and there is no need for human *in vivo* data) for immediate release BCS class 1 drug products and an extension of the scheme to some class 3 drugs has been proposed (Tsume and Amidon, 2010).

Note that the BCS categories take account of drug dose and hence highly potent (and so low dose) drugs with low aqueous solubilities can still be categorised in classes 1 and 3. The USP and PhEur provide definitions of solubility based on concentration alone (Table 4.2 – the two pharmacopoeia use the same definitions with the exception of 'practically insoluble' which is absent from the PhEur).

In any event, determining the solubility of a new drug substance is an essential first step in assessing its likely developability. Experimental measurement of solubility is tricky, being time consuming as well as analytically challenging if the compound dissolves to a very low extent and/or undergoes hydrolysis. Furthermore, in early preformulation only small quantities (<50 mg) of a drug substance may exist and neither its purity nor polymorphic form may be assured. Computer modelling approaches (such as MOLPRO, ChemDBsoft, TimTec LLC or ALOGPS, Virtual Computational Chemistry Laboratory) have been developed that allow prediction of solubility from chemical structure, but such results are inherently approximate and so are best validated with experimental data.

Initial formulations, used for obtaining toxicity and bioavailability data in animal models, will need to be liquids (either solutions or micronised drug

Table 4.2   Table of USP and PhEur terms for describing the solubility of drugs.

Descriptive term	Parts solvent to 1 part solute	Solubility range (mg mL^{-1})
Very soluble	Less than 1	>1000
Freely soluble	1–10	100–1000
Soluble	10–30	33–100
Sparingly soluble	30–100	10–33
Slightly soluble	100–1000	1–10
Very slightly soluble	1000–10 000	0.1–1
Practically insoluble	More than 10 000	<0.1

suspensions) for gavage or intravenous delivery and solubility greater than 1 mg mL^{-1} is usually acceptable. For the drug product, assuming oral delivery in a solid form, solubility above 10 mg mL^{-1} is preferable (Kaplan, 1972), although these limits are arbitrary and may be reduced if the drug substance is highly potent. If the solubility of the drug substance is found to be below 1 mg mL^{-1} then salt formation (Chapter 6) is an option. Where solubility cannot be manipulated with salt formation, a novel dosage form design will be required, although it must be stressed that the simplest, most robust drug product formulations are likely to have the greatest chance of reaching the market.

High solubility does not of itself indicate fast dissolution, since solubility is a position of equilibrium and dissolution is the rate at which equilibrium is established. This means a drug substance with good solubility may not necessarily have satisfactory bioavailability. It has been suggested that absorption will be unimpeded if a drug substance has an intrinsic dissolution rate (IDR) (see Chapter 5) greater than 1 mg cm^{-2} min^{-1} but is likely to be limited if the IDR is less than 0.0016 mg cm^{-2} min^{-1} (Kaplan, 1972).

4.2 Intrinsic solubility

Solid drug substances are held together in a three-dimensional array as a result of intermolecular bonds. The strength of the bonds (the crystal lattice energy) governs many of the physical attributes of the macroscopic crystal (such as hardness, melting point and so on). As the temperature of a solid increases so does the energy it contains. From everyday experience, when the temperature reaches a certain critical level (the melting temperature range) the solid undergoes a phase transition to form a liquid (usually, although in some cases compounds transform directly to a gas – *sublimation* – the barbiturates are a series of drugs that sublime). From a thermodynamic perspective, the internal energy of the solid becomes greater than the crystal lattice energy and so the intermolecular bonds are broken. The amount of energy (heat) required to melt the solid is known as the enthalpy of fusion ($\Delta_f H$) and is usually expressed in units of kJ mol^{-1} or kJ g^{-1}.

If the molecules are able to align in different repeating patterns (the phenomenon of polymorphism, explained further in Chapter 7) then it is highly likely that the strength of the intermolecular bonds, and hence crystal lattice energy, will change. It follows that two polymorphs of the same drug will have different melting temperature ranges and enthalpies of fusion and, as a consequence, a range of other different physical properties, including dissolution rate and thermodynamic solubility.[1] Knowledge of the melting range

---

[1]Thermodynamic solubility relates to the position of equilibrium formed between solid and dissolved solute. Kinetic solubility is the concentration achieved in buffer following addition of a stock solution of solute in a nonaqueous solvent, typically DMSO. This text relates to thermodynamic solubility.

and enthalpy of fusion provides fundamental insight into the character of different physical forms, which is why these parameters are often among the first to be quantified during preformulation (and both, incidentally, can be determined with differential scanning calorimetry, or DSC).

What does melting have to do with dissolution, as most drugs are evidently not administered in the molten state? Well, dissolution is itself a phase transition, with the drug transforming from a solid matrix to a dissolved solute. In order for this phase transition to progress, solute–solute interactions must be overcome (in effect, the solid melts[2]), while solvent–solvent interactions must be replaced with solute–solvent interactions (the solute molecules become *solvated*). Assuming that an excess of solid was added initially, and so solid material is always present, a position of equilibrium will be established between solid and dissolved drug:

$$\text{Drug}_{(s)} \rightleftharpoons \text{Drug}_{(l)} \rightleftharpoons \text{Drug}_{(aq)} \tag{4.1}$$

The concentration of drug dissolved at this point is known as the *equilibrium solubility* (usually the term solubility is used alone) and the solution is *saturated*. If the drug has an ionisable group then the equilibrium solubility of the un-ionised form is called the *intrinsic solubility* (S_o). This is important, because if the drug is ionisable it will ionise to a greater or lesser extent with solution pH and this will affect the observed solubility, as discussed in Chapter 2. If the structure of the compound is known, then it should be clear whether solubility will exhibit a dependence upon pH. If the structure is not known then measuring solubility over a range of pH will show whether an ionisable moiety is present (although care must be taken when selecting the buffer to ensure salts are not accidentally formed). If the compound has an ionisable group, then modification of solubility by preparation of a salt is a possible formulation strategy (see Chapter 6).

From a thermodynamic perspective, the energy input required in order to break any solute–solute interactions must be equal to the enthalpy of fusion required to melt the solid (since the same interactions must be overcome). Unlike melting, however, in the case of dissolution there is an additional change in enthalpy where solvent–solvent intermolecular interactions are broken and solute–solvent interactions are formed, termed the enthalpy of mixing ($\Delta_{mix}H$). The net enthalpy of solution ($\Delta_{sol}H$) is then the sum of the enthalpies of fusion and mixing:

$$\Delta_{sol}H = \Delta_f H + \Delta_{mix}H \tag{4.2}$$

---

[2]This is the basis of much debate, but for the purpose of the following discussion it is the most convenient theoretical construct.

The overall process of dissolution can be represented more simply as

$$Drug_{(s)} \rightleftharpoons Drug_{(aq)} \tag{4.3}$$

In this case the equilibrium constant (K) can be written as

$$K_{sol} = \frac{a_{(aq)}}{a_{(s)}} \tag{4.4}$$

where a denotes the activity of the drug in each phase. Since the activity of a solid is defined as unity and activity approximates to concentration (saturation in this case) in dilute solution then

$$K_s = S_o = x_2 \tag{4.5}$$

where x_2 denotes the saturated concentration of drug in mole fraction units (x_1 being the mole fraction of the solvent). It should now be possible to see from Equation (4.5) the reason for the term equilibrium solubility noted earlier.

It appears from Equation (4.2) that the crystal lattice energy might affect solubility. It also seems from Equation (4.1) that there should be an effect of temperature on solubility, since the position of equilibrium will change. Both of these effects can be explored further through the concept of *ideal solubility*.

Summary box 4.1

- Solubility is the maximum concentration of a given solute that can be attained in a given solvent.

- Solids transition to solution by dissolution.

- Thermodynamic solubility is a position of equilibrium.

- Dissolution governs the rate at which solubility is achieved.

- As a general rule, solubility below 1 mg mL^{-1} is likely to hinder development while solubility above 10 mg mL^{-1} is acceptable.

4.2.1 Ideal solubility

In the special case where the enthalpy of any solute–solvent interaction is equal to the enthalpy of any solvent–solvent interaction then solvation of the solute may occur with no change in enthalpy (i.e. $\Delta_{mix}H = 0$) and dissolution is said to be *ideal*. Formation of an ideal solution also occurs with the following change in entropy ($\Delta_{sol}S$):

$$\Delta_{sol}S = -R(x_1 \ln x_1 + x_2 \ln x_2) \tag{4.6}$$

where R is the universal gas constant ($8.314 \text{ J K}^{-1}\text{mol}^{-1}$). Ideal dissolution (although unlikely, because the solute and solvent molecules would need to possess identical properties, such as size, shape and chemical nature) leads to ideal solubility and is an interesting theoretical position because it can be described in thermodynamic terms, which allows calculation of the dependence of solubility on temperature.

From Equation (4.2) if $\Delta_{mix}H = 0$ then $\Delta_f H$ is equal to $\Delta_{sol}H$ (note that since $\Delta_f H$ must be positive, i.e. endothermic, $\Delta_{sol}H$ must also be positive for ideal dissolution). For a process to occur spontaneously the Gibbs free energy (ΔG) must be negative. The familiar thermodynamic relationship for dissolution is

$$\Delta_{sol}G = \Delta_{sol}H - T\Delta_{sol}S \tag{4.7}$$

where T is absolute temperature. $\Delta_{sol}G$ is most likely to be negative when $\Delta_{sol}H$ is negative but, as noted above, $\Delta_{sol}H$ is frequently positive for dissolution and must be so when dissolution is ideal. This means that for dissolution to occur spontaneously the driving force can *only* be a significant increase in entropy. Since the mole fractions of both solvent and solute must be less than 1, the logarithmic terms in Equation (4.6) must be negative and so the change in entropy must be positive, which agrees with this argument.

Equation (4.5) shows that solubility is a position of equilibrium and thus has the attributes of an equilibrium constant. If so, it is possible to apply the van't Hoff equation, yielding

$$\frac{d \ln x_2}{dT} = \frac{\Delta_f H}{RT^2} \tag{4.8}$$

Assuming that $\Delta_f H$ is independent of temperature (in effect, ignoring any change in heat capacity) then integrating Equation (4.8) from T_m to T results in

$$\ln x_2 = \frac{-\Delta_f H}{RT} + \frac{\Delta_f H}{RT_m} \tag{4.9}$$

where T_m is the melting temperature (usually the extrapolated onset temperature of melting, as melting typically occurs over a range of 1–2 °C) of the pure drug substance and T is the experimental temperature. Equation (4.9) is very useful, since it allows prediction of ideal solubility at any particular temperature so long as the melting temperature and enthalpy of fusion of the pure drug substance are known (both are easily determined from a DSC measurement).

Example 4.1   The melting temperature of aspirin is $137\,°C$ and its enthalpy of fusion at the melting temperature is $29.80\ kJ\ mol^{-1}$. What is the ideal solubility of aspirin at $25\,°C$?
    Applying Equation (4.9),

$$\ln x_2 = \frac{-29800}{8.314 \times 298} + \frac{29800}{8.314 \times 410} = -3.286$$

$$x_2 = 0.037$$

Of course, real solutions invariably do not show ideal behaviour, because the assumptions made above that $\Delta_{mix}H = 0$ and that effects of heat capacity can be ignored are not always valid. A negative (exothermic) enthalpy of mixing increases solubility while a positive (endothermic) enthalpy of mixing reduces solubility. Tables 4.3 and 4.4 list the experimentally measured solubilities for aspirin and paracetamol in a range of solvents. Note that solubility in water is by far the lowest of any of the solvents shown, while solubilities in THF and methanol approach ideality in the case of aspirin and exceed ideality in the case of paracetamol. This explains why THF and methanol are frequently used as solvents to aid dissolution of poorly soluble drug substances during preformulation assay development. Methanol, as noted in Chapter 2, is also

Table 4.3   Ideal (calculated) solubility for aspirin compared with experimentally determined solubilities in a range of solvents (at 25 °C, assuming MP 137.23 °C, $\Delta_f H$ 29.8 kJ mol^{-1}) (data from the RSC Open Notebook Science Challenge).

Solvent	Solubility (mole fraction)
Ideal (calculated)	0.037
THF	0.036
Methanol	0.025
Ethanol	0.023
Acetone	0.018
Chloroform	0.015
1-Propanol	0.011
Acetonitrile	0.006
Water	0.000 45

Table 4.4   Ideal (calculated) solubility for paracetamol compared
with experimentally determined solubilities in a range of solvents
(at 30 °C, assuming MP 170 °C, $\Delta_f H$ 27.6 kJ mol^{-1}) (data from
Granberg and Rasmuson (1999)).

Solvent	Solubility (mole fraction)
Ideal (calculated)	0.031
Diethylamine	0.389
Methanol	0.073
THF	0.069
Ethanol	0.066
1-Propanol	0.051
Acetone	0.041
Acetonitrile	0.009
Water	0.002

commonly used as a co-solvent for determination of pK_a for poorly soluble
compounds.

The reason for nonideal solubility in so many solvents, water in particular, is because of significant intermolecular interactions resulting from chemical structure and properties. The three primary properties are the *dipole
moment, dielectric constant* and capacity for forming *hydrogen bonds*.

A molecule has a dipole when there is a localised net positive charge at
one end and a localised net negative charge at the other; such molecules are
said to be polar. Water is an example of a polar molecule. Drug substances
that have dipoles or a dipolar character are generally more soluble in polar
solvents. This is an important point to remember during later chapters.[3]

Dielectric properties are related to the capacity of a molecule to store a
charge and are quantified by a dielectric constant. Polar solvents may induce
a dipole in a dissolved solute, which will increase solubility. The dielectric
constants of a number of commonly used pharmaceutical solvents are given
in Table 4.5. It can be seen that water has a high dielectric constant (78.5)
relative to methanol (31.5) even though both are considered to be polar
solvents.

Hydrogen bonding occurs when electronegative atoms (such as oxygen) come into close proximity with hydrogen atoms; electrons are pulled
towards the electronegative atom creating a reasonably strong force of interaction. A drug substance that has a functional group capable of hydrogen
bonding with water (such as –OH, –NH or –SH) should have increased
aqueous solubility.

---

[3]This leads to the joke, 'Why did the white bear dissolve in water? Because it was polar.'

Table 4.5 Dielectric constants of some common pharmaceutical solvents at 25 °C.

Solvent	Dielectric constant
Water	78.5
Glycerin	40.1
Methanol	31.5
Ethanol	24.3
Acetone	19.1
Benzyl alcohol	13.1
Phenol	9.7
Ether	4.3
Ethyl acetate	3.0

4.2.2 Solubility as a function of temperature

Since $\Delta_f H$ must be positive, Equation (4.9) suggests that the solubility of a drug should increase with an increase in temperature. Generally, this agrees with everyday experience (Figure 4.1 shows the relationship between solubility and temperature for paracetamol in three solvents), but there are some drug substances (9,3″-diacetylmidecamycin, for instance; Sato *et al.*, 1981) for which solubility decreases with increasing temperature. This is because the

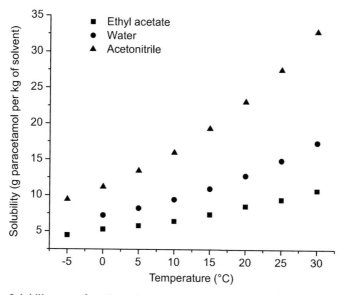

Figure 4.1 Solubility as a function of temperature for paracetamol in three solvents (data from Granberg and Rasmuson (1999)).

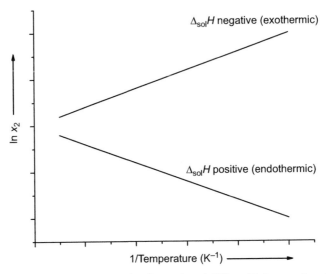

Figure 4.2 A schematic plot showing the change in solubility with temperature for drugs with endothermic and exothermic enthalpies of solution.

assumption was made in deriving Equation (4.9) that $\Delta_f H$ was equal to $\Delta_{sol} H$. However, as noted above and as demonstrated by the data in Tables 4.3 and 4.4, $\Delta_{mix} H$ is frequently not zero and hence the equality assumed in Equation (4.2) cannot be true. In cases where $\Delta_{sol} H$ is negative (i.e. the enthalpy of solution is exothermic) solubility will decrease with increasing temperature. Schematically, these effects are shown in Figure 4.2.

It can be seen from Equation (4.9) that the enthalpy of fusion should be determinable by experimentally measuring the solubility of a drug at a number of temperatures (since a plot of ln x_2 versus $1/T$ should be linear and of slope $-\Delta_f H/R$) (Prankerd, 1992). Examples of data plotted in this way, for paracetamol, allopurinol and budesonide, are given in Figure 4.3. While such plots are frequently found to be linear, they are usually plotted over a very narrow temperature range and the enthalpy of fusion so calculated is rarely ideal, although it can be considered to be an approximate heat of solution.

4.2.3 Solubility and physical form

If two drug substances have the same enthalpy of fusion, then Equation (4.9) predicts that the drug substance with the lower melting point will have the greater solubility. Similarly, if two drug substances have the same melting point, solubility is dependent upon the enthalpy of fusion, the form with the lower enthalpy of fusion exhibiting greater solubility.

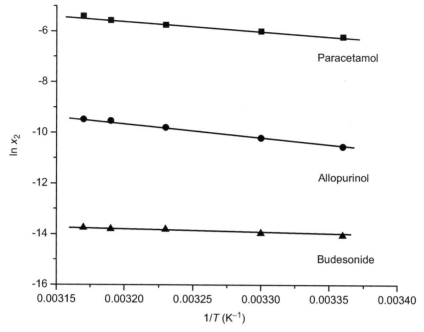

Figure 4.3   A plot of ln x_2 versus $1/T$ for three drugs in water (solubility data from Mota *et al.* (2009)).

Of more interest pharmaceutically is the effect of polymorphism, where the same drug substance can exist in multiple solid forms, each with a different crystal lattice. In this case the chemistry of the molecule is the same but both the enthalpy of fusion and melting point will be different, as a result of differences in the crystal lattice energies. Each polymorph must have a unique solubility (with the exception of an enantiotropic pair at their transition temperature). Usually the stable polymorph has the highest melting point and greatest enthalpy of fusion and so, from Equation (4.9), the lowest solubility. Any metastable forms will, by definition, have lower melting points and enthalpies of fusion and so greater solubilities. The form with the lowest melting point and enthalpy of fusion will exhibit the greatest solubility, which in the limit is the amorphous form (no melting point or enthalpy of fusion).

An extremely important point to note here is that solubility is defined as the equilibrium between the dissolved solute and the solid form. Thus, if a saturated solution is prepared from a metastable polymorph or amorphous form and the excess solid is removed by filtration, the solution remaining can be considered to be supersaturated with respect to the stable form. Ultimately the stable form will precipitate as the system establishes a new position of equilibrium (Figure 4.4).

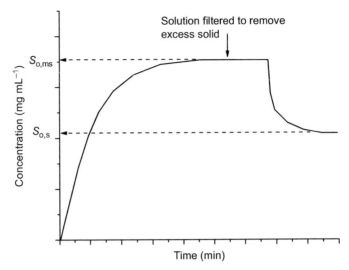

Figure 4.4 Concentration versus time profile for dissolution of a metastable (ms) form of a drug. The system is in equilibrium until excess drug is removed by filtration, after which the solution is supersaturated with respect to the stable (s) form. Subsequently the stable form precipitates and a new position of equilibrium is reached.

Study question 4.1   What is the risk of choosing a metastable form of a drug substance for development?

Summary box 4.2

- Different polymorphs of a drug will have different solubilities.

- If the excess solid is removed after equilibrium has been attained and the polymorph dissolved was metastable, the solution remaining can be considered supersaturated with respect to the stable form and crystallisation will eventually occur.

- Solubility generally increases with temperature.

- Knowledge of the enthalpy of fusion and the melting temperature allows calculation of ideal solubility.

- Solubility in water is rarely ideal, because of solvent–solute and solvent–solvent interactions, but often approaches ideality in methanol and THF, which is why these solvents are used to aid dissolution of poorly soluble compounds.

4.2.4 Measurement of intrinsic solubility

Initially, solubility should be determined in 0.1M HCl, 0.1M NaOH and water. Saturated solutions can be prepared by adding an excess of solid to a small volume of solvent, agitating with time and then filtering. A simple UV assay (if available) will allow rapid determination of concentrations. Measurements should be recorded until the concentration remains constant and at a maximum. Care should be taken to ensure the drug substance is not degrading if hydrolysis or photolysis are potential reaction pathways and also that temperature does not fluctuate. If the solubilities are the same in the three solvents then the drug substance does not have an ionisable group. If solubility is highest in acid then the molecule is a weak base and if solubility is highest in alkali the molecule is a weak acid. Zwitterionic compounds will not show a discernable trend. In any of these three cases the molecule will have an ionisation constant and should be amenable to salt formation, if solubility manipulation is necessary. If time permits, solubility can be measured at a number of temperatures, because if the melting point of the pure compound is known then the enthalpy of fusion can be estimated from Equation (4.9):

- 4 °C: the reduced temperature minimises the rate of hydrolysis (if applicable) and the density of water is at its greatest, which presents the greatest challenge to solubility.

- 25 °C: standard room temperature.

- 37 °C: body temperature and so an indication of solubility *in vivo*.

If the aim of the preformulation screen is to understand solubility *in vivo* then solubility in gastrointestinal (or simulated gastrointestinal) fluids should be determined. Assuming oral delivery, typical media would include simulated gastric fluid (SGF), fed state simulated gastric fluid (FeSSGF), fed state simulated intestinal fluid (FeSSIF) or fasted state simulated intestinal fluid (FaSSIF). Gastrointestinal fluids tend to have higher ionic strengths and hence the risk of salting out via the common ion effect is greater.

One potential issue at this stage is the polymorphic form of the drug. It may well be the case that any impurities present have led to crystallisation of a metastable form. It is a good idea to use X-ray powder diffraction (XRPD) to determine the polymorph of the excess solid filtered from the solubility experiments to ensure that there has been no form change to a stable polymorph or that a hydrate has not formed (since both forms typically have lower solubilities).

Another potential issue is the chemical purity of the sample. If the drug substance is pure then its phase–solubility diagram should appear as drawn

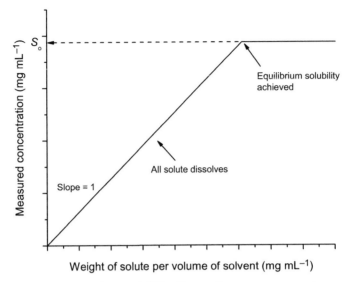

Figure 4.5   Phase–solubility diagram for a pure compound.

in Figure 4.5. Initially all of the drug substance added to solvent dissolves and the gradient of the line should be unity. When saturation is achieved addition of a further drug substance does not result in an increase in concentration and the gradient reduces to zero. When a single impurity is present, the phase–solubility diagram appears as drawn in Figure 4.6. From the origin to point A both components dissolve. At point A the first compound has reached its solubility. The line AB represents the continued dissolution of the second compound. At point B the second compound reaches its solubility and the

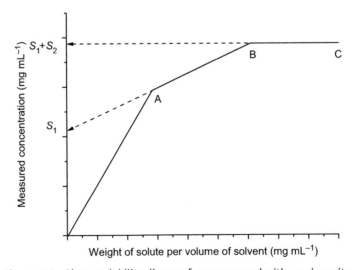

Figure 4.6   Phase–solubility diagram for a compound with one impurity.

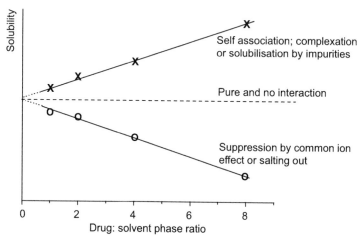

Figure 4.7 Effect of the drug:solvent phase ratio when the drug is impure (redrawn from Wells, 1988, with permission from John Wiley & Sons, Inc.).

gradient of the line BC is zero. The solubility of the first compound can be determined by extrapolation of line AB to the y axis. The solubility of the second compound is the difference between the solubility at CD and the y intercept of the extrapolated line AB. The same principle applies if further impurities are present.

An alternative experiment is to prepare four solutions of drug substance with different phase ratios of drug substance to solvent (say 3, 6, 12 and 24 mg in 3 mL), measure the solubility of each and then extrapolate the data to a phase ratio of zero (Figure 4.7; Wells, 1988). If the drug substance is pure then the solubility should be independent of phase ratio. If the impurity acts to increase solubility (say by self-association, complexation or solubilisation) then the gradient of the line will be positive, whereas if the impurity acts to suppress solubility (say by the common ion effect) then the gradient of the line will be negative.

The purity of a sample may also be checked with DSC (Marti, 1972), since the presence of (even minor amounts of) an impurity will lower and broaden the melting point of a material (in accordance with van't Hoff's law of freezing point depression). Qualitatively, if the melting endotherm recorded with DSC is very broad then the sample is likely to be impure (Figure 4.8). Of course, the sample may itself degrade during heating, especially if a slow heating rate is used, so the presence of a broad melting endotherm is not necessarily indicative of an impure sample. One way to minimise this effect is to use fast heating rates. If the melting point and heat of fusion of the pure drug substance are known, then purity of an impure sample can be quantified by analysis of DSC data (see, for instance, the method in ASTM E928-08, 2008).

Briefly, analysis requires determination of the fraction of sample melted as a function of temperature. This is easily achieved by recognition of the fact

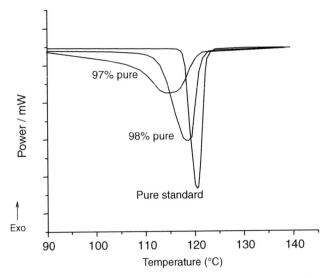

Figure 4.8 DSC thermal traces for benzoic acid of varying purity.

that integration of the peak area of melting (note, with respect to *time* rather than temperature) gives the total enthalpy of fusion (Q). Partial integration of the melting endotherm to any particular temperature must therefore give a smaller enthalpy (q_T). The fraction of material melted at any temperature (F_T) is given by

$$F_T = \frac{q_T}{Q} \qquad (4.10)$$

Time and temperature are interchangeable, since linear heating rates are used in DSC, and so a table of F_t values as a function of temperature can be created. The van't Hoff equation (Equation (4.11) – see Marti, 1972, for the derivation) predicts that a plot of $1/F_t$ versus temperature should be a straight line of slope $-RT_m^2 x_2 / \Delta H$

$$T = T_m - \frac{RT_m^2 x_2}{\Delta H} \frac{1}{F_t} \qquad (4.11)$$

Figure 4.9 shows the melting endotherm for indomethacin recorded with DSC and Figure 4.10 shows the corresponding van't Hoff plot. In practice the method often results in van't Hoff plots that are significantly nonlinear, a result of many of the assumptions inherent in the van't Hoff analysis not being true. The main assumptions are:

- The melting point is the triple point.

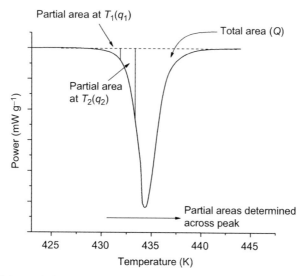

Figure 4.9   Melting endotherm for indomethacin recorded by DSC and the integration of partial areas to allow calculation of sample purity.

- Thermal equilibrium must be maintained during melting, so slow heating rates must be used (and so the possibility of considerable degradation occurring during heating should not be excluded).

- The components must form eutectic mixtures and any impurities must dissolve in the main component to form an ideal solution.

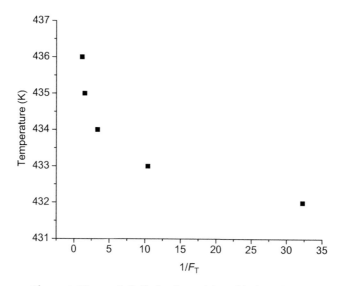

Figure 4.10   van't Hoff plot for melting of indomethacin.

- The enthalpy of fusion is constant with temperature.

- The mole fraction of the impurity remains constant (i.e. further degradation does not occur during heating, nor is there loss of any volatile component).

In an excellent review, van Dooren and Müller (1984) discuss these limitations and review various methods of 'linearising' the van't Hoff plot. One method is that of Sondack (1972), in which a correction factor is applied to the fractional areas calculated with Equation (4.10). The correction factor (X) is determined by selecting three pairs of temperatures and partial areas:

$$X = \frac{\dfrac{T_3 - T_2}{T_2 - T_1}a_3 - \dfrac{a_3 - a_2}{a_2 - a_1}a_1}{\dfrac{a_3 - a_2}{a_2 - a_1} - \dfrac{T_3 - T_2}{T_2 - T_1}} \qquad (4.12)$$

F_T values are then corrected as

$$F_T = \frac{q_T + X}{Q + X} \qquad (4.13)$$

Applying Sondack's correction to the data for indomethacin results in a corrected, and linear, van't Hoff plot (Figure 4.11).

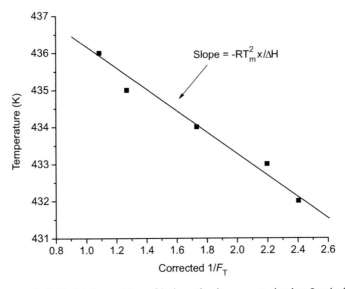

Figure 4.11   van't Hoff plot for melting of indomethacin, corrected using Sondack's method.

4.2.5 Calculation of pK_a from solubility data

If the intrinsic solubility has been determined, measurement of solubility at a pH where the compound is partially ionised can allow estimation of pK_a from the Henderson–Hasselbalch equations derived in Chapter 2, although significant errors may occur.

Example 4.2 What is the pK_a of the weak base chlordiazepoxide given the following solubility data: $S_o = 2$ mg mL^{-1}, S_t at pH $4 = 14.6$ mg mL^{-1}, S_t at pH $6 = 2.13$ mg mL^{-1}?

For a weak base:

$$pK_a = pH + \log\frac{S_t - S_o}{S_o}$$

At pH 4:

$$pK_a = 4 + \log\frac{14.6 - 2}{2} = 4.799$$

At pH 6:

$$pK_a = 6 + \log\frac{2.13 - 2}{2} = 4.813$$

The literature value is 4.8.

4.3 Summary

High aqueous solubility is obviously a preferable quality of a drug substance, but low solubility is not necessarily a barrier to successful development, since the formulation may be designed to help aid solubility or the drug may be highly potent (and so only low doses are required). Knowledge of the melting point and enthalpy of fusion allows calculation of the ideal solubility; frequently aqueous solubility is considerably nonideal, primarily because of solute–solvent interactions. Solubility in THF or methanol is usually much nearer to ideal and so these solvents are used to aid solubilisation in early assays. Different physical forms of a compound will have different solubilities and dissolution rates, so selecting a particular physical form to ensure bioavailability is one strategy for development.

References

ASTM standard E928-08 (2008) Standard test method for determination of purity by differential scanning calorimetry. ASTM International, West Conshohocken, PA. DOI:10.1520/E0928-08.

Amidon, G.L., Lennernäs, H., Shah, V.P. and Crison, J.R. (1995) A theoretical basis for a biopharmaceutical drug classification: The correlation of *in vitro* drug product dissolution and *in vivo* bioavailability. *Pharm. Res.*, **12**, 413–420.

Granberg, R.A. and Rasmuson, A.C. (1999) Solubility of paracetamol in pure solvents. *J. Chem. Engng Data*, **44**, 1391–1395.

Kaplan, S.A. (1972) Biopharmaceutical considerations in drug formulation design and evaluation. *Drug Met. Rev.*, **1**, 15–34.

Kennedy, T. (1997) Managing the drug discovery/development interface. *Drug Disc. Today*, **2**, 436–444.

Marti, E.E. (1972) Purity determination by differential scanning calorimetry. *Thermochim. Acta*, **5**, 173–220.

Mota, F.L., Carneiro, A.P., Queimada, A.J. *et al.* (2009) Temperature and solvent effects in the solubility of some pharmaceutical compounds: measurements and modeling. *Eur. J. Pharm. Sci.*, **37**, 499–507.

Prankerd, R.J. (1992) Solid state properties of drugs. I. Estimation of heat capacities for fusion and thermodynamic functions for solution from aqueous solubility–temperature dependence measurements. *Int. J. Pharm.*, **84**, 233–244.

Sato, T., Okada, A., Sekiguchi, K. and Tsuda, Y. (1981) Difference in physic-chemical properties between crystalline and non-crystalline 9,3″-diacetylmidecamycin. *Chem. Pharm. Bull.*, **29**, 2675–2682.

Sondack, D.L. (1972) Simple equation for linearization of data in differential scanning calorimetric purity determinations. *Anal. Chem.*, **44**, 888.

Stegemann, S., Leveiller, F., Franchi, D. *et al.* (2007) When poor solubility becomes an issue: from early stage to proof of concept. *Eur. J. Pharm. Sci.*, **31**, 249–261.

Tsume, Y. and Amidon, G.L. (2010) The biowaiver extension for BCS class III drugs: the effect of dissolution rate on the bioequivalence of BCS class III immediate-release drugs predicted by computer simulation. *Molec. Pharm.*, **7**, 1235–1243.

van Dooren, A.A. and Müller, B.W. (1984) Purity determinations of drugs with differential scanning calorimetry (DSC) – a critical review. *Int. J. Pharm.*, **20**, 217–233.

Wells, J.I. (1988) *Pharmaceutical Preformulation. The Physicochemical Properties of Drug Substances*, John Wiley & Sons, Ltd, Chichester. ISBN 0-470-21114-8.

Answer to study question

4.1 Formulating any drug in a metastable form will involve an element of risk, the risk being that the stable form will appear during storage or post dissolution; in either case the maximum aqueous concentration achievable will be reduced, with a concomitant reduction in bioavailability.

Additional self-study questions and answers

4.1 The melting temperature of paracetamol is 170 °C and its heat of fusion at the melting temperature is 27.6 kJ mol^{-1}. What is the ideal solubility of paracetamol at 30 °C?

Applying Equation (4.9),

$$\ln x_2 = \frac{-27600}{8.314 \times 303} + \frac{27600}{8.314 \times 443} = -3.462$$

$$x_2 = 0.0314$$

4.2 The melting temperature of budesonide is 261 °C and its heat of fusion at the melting temperature is 34.7 kJ mol^{-1}. What is the ideal solubility of budesonide at 25 °C?

Applying Equation (4.9),

$$\ln x_2 = \frac{-34700}{8.314 \times 298} + \frac{34700}{8.314 \times 534} = -6.190$$

$$x_2 = 0.00205$$

5
Dissolution

5.1 Introduction

The concepts of ionisation and solubility are equilibria (i.e. they represent a steady-state condition reached after completion of a process). Dissolution, conversely, is defined by a rate and so is a kinetic process. Fast dissolution is often linked with good bioavailability and so is a desirable quality in an immediate release product.

5.2 Models of dissolution

Early work by Noyes and Whitney (1897) into the dissolution of two model compounds (lead chloride and benzoic acid) led to the formulation of a general rate law that showed the dissolution rate to be proportional to the difference in solution concentration (C) relative to solubility:

$$\frac{dC}{dt} = k(S_t - C) \tag{5.1}$$

Additional work by various researchers elucidated the experimental factors that contributed to the proportionality constant k in Equation (5.1) (Brunner and Tolloczko, 1900; Brunner, 1904; Nernst 1904), leading to the form of the Noyes–Whitney equation that is still used today:

$$\frac{dm}{dt} = \frac{DA}{h}(S_t - C) \tag{5.2}$$

where m is the mass of drug, D is the diffusion coefficient, A is the surface area of dissolving solid, h is the thickness of the unstirred water layer (the

Essentials of Pharmaceutical Preformulation, First Edition. Simon Gaisford and Mark Saunders.
© 2013 John Wiley & Sons, Ltd. Published 2013 by John Wiley & Sons, Ltd.

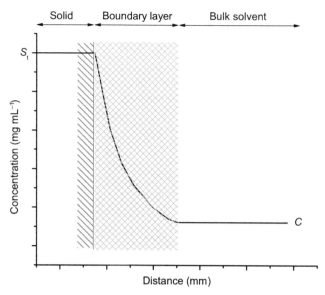

Figure 5.1 A schematic representation of the boundary layer adjacent to the surface of a dissolving solid and the change in concentration of solute across it.

boundary layer) surrounding the dissolving solid and C is the concentration of drug in bulk solvent. An alternative version of Equation (5.2) is the Noyes–Whitney–Nernst–Brunner (NWNB) equation, which describes the change in concentration of dissolved solid with time:

$$\frac{dC}{dt} = \frac{DA}{Vh}(S_t - C) \tag{5.3}$$

where V is the volume of bulk solvent. In either case, the model assumes that any drug molecule that dissolves at the surface of the solid must then diffuse through a stagnant layer of saturated drug solution surrounding the solid before it moves into the bulk solvent (Figure 5.1). Several other notable models for dissolution have been published, including the interfacial barrier model (Wilderman, 1909) and the Danckwerts' model (in which a constant stream of macroscopic packets of solvent arrive at the solid surface to absorb solute molecules and transport them to the bulk solvent; Danckwerts, 1951), but the NWNB model is still the model most commonly used.

5.3 Dissolution testing

Concern over the importance of the relationship between the dissolution rate and bioavailability led to the introduction of rotating basket dissolution tests (i.e. a uniform test to measure the dissolution rate in a defined medium) to the US pharmacopoeia in 1970 (although a rotating dissolution method for

Table 5.1 The number of USP monographs for immediate release oral dosage forms requiring dissolution tests (reproduced from Dokoumetzidis and Macheras, Copyright 2006, with permission from Elsevier).

USP edition/year	Monographs for immediate release dosage forms requiring dissolution testing
USP 18-NF 13/1970	6
USP 19-NF 14/1975	12
USP 20-NF 15/1980	60
USP 21-NF 16/1985	400
USP 22-NF 17/1990	462
USP 23-NF 18/1995	501
USP 24-NF 19/2000	552
USP 29-NF 20/2000	619

extended release products was introduced in 1958). In an excellent review of the history of dissolution testing Dokoumetzidis and Macheras (2006) chart the rise in the number of monographs in the US pharmacopoeia that require dissolution tests (Table 5.1). All pharmacopoeias specify dissolution apparatus that should be used for measuring dissolution data (Table 5.2).

Apparatuses 1 and 2 are similar in arrangement, varying only in the stirrer design (the sample is contained within a rotating basket, apparatus 1, or stirred with a paddle, apparatus 2; Figure 5.2). The dissolution medium (900 or 1000 mL) is contained in a round-bottomed vessel housed in a thermostatted water bath. Both methods allow the pH of the dissolution medium to be changed during the test (although this is slightly easier with apparatus 1 in cases where the drug product is contained within the basket), mimicking the change in pH along the GI tract. One problem with apparatus 2 is that the drug product might float, in which case a 'sinker' (typically a small piece of coiled, nonreactive wire) can be attached to the drug product.

Table 5.2 Pharmacopoeial dissolution apparatus.

Apparatus	Type	Volume	Application
1	Basket	900/1000 mL	Capsules, beads, delayed release or floating products
2	Paddle	900/1000 mL	Tablets, capsules, beads, delayed release products
3	Reciprocating cylinder	200–250 mL	Tablets, beads, controlled release products
4	Flow-through cell	Open loop – unlimited volume	Microparticles, implants, suppositories, stents
5	Paddle over disc	900 mL	Transdermal patches
6	Rotating cylinder	900 mL	Transdermal patches

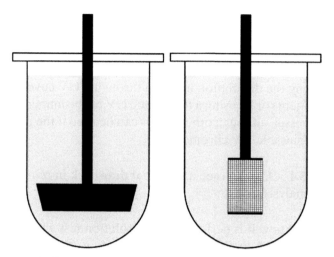

Figure 5.2 Dissolution apparatus 1 (basket, right) and apparatus 2 (paddle, left).

The hydrodynamics of the medium will vary with stirring rate, which may also affect dissolution. In particular, slow stirring speeds in apparatus 2 can lead to a 'coning' effect, where the dissolving compact forms a cone at the bottom of the vessel. This effect has been shown to reduce the dissolution rate for immediate release products that contain a large proportion of insoluble excipients or products containing BCS class II or IV poorly soluble drug substances (Klein, 2006). Figure 5.3 shows the dissolution

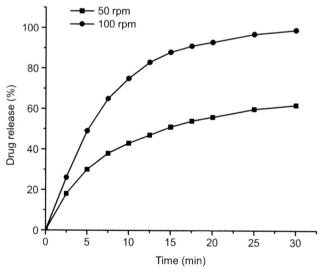

Figure 5.3 Dissolution curves for indomethacin tablets stirred at 50 and 100 rpm (data redrawn from Klein (2006)).

profiles for indomethacin tablets at two stirring speeds where this effect can be seen.

Drug concentrations are most conveniently determined with UV spectrophotometry, either by removing aliquots of solution at regular time intervals, by circulating the dissolution medium through a UV cuvette and taking in-line measurements or by using a fibre-optic UV probe immersed directly in the dissolution vessel, although other assays can be used if the drug substance does not have a suitable UV chromophore.

Study question 5.1   Can you see a potential drawback in removing aliquots of solution for analysis?

Irrespective of how it is performed, a dissolution test will result in a plot of concentration versus time (Figure 5.4). Equation (5.3) predicts that the rate of change of concentration with time will decrease as the concentration of solute increases; as depicted in Figure 5.4, the rate of dissolution (given by the gradient of the line) starts at a maximum and asymptotically approaches zero as the concentration approaches the solubility. A further complication is that the surface area of the dissolving solid will also change with time (initially there may be a large increase if, for instance, a compact disintegrates, followed by a decrease to zero) and this will affect the dissolution rate. It is thus difficult to define a dissolution rate for a given material since the rate is constantly decreasing. A general convention is to construct dissolution experiments such that the final concentration attainable, if all the material dissolves,

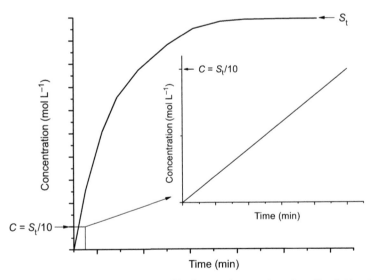

Figure 5.4 Concentration versus time profile for a solute undergoing dissolution following addition to water and (inset) the linear rate region when $C = S_t/10$ (i.e. sink conditions).

is never greater that 10% of the equilibrium solubility. The inset graph of Figure 5.4 shows the dissolution data plotted over this 10% region and it can be seen that the gradient of the line (and hence the dissolution rate) is essentially constant. When constructed in this way the dissolution experiment has been performed under *sink conditions*.

Study question 5.2   Would experimentally measured dissolution data recorded under sink conditions show such linearity?

It is extremely important to understand that dissolution testing was introduced primarily as an *assay* to measure batch-to-batch variability on a production scale and that the dissolution data produced were initially used to set batch release standards. In this context, whether the conditions of the dissolution test match those likely to be experienced by the drug product *in vivo* is irrelevant. However, this principle is often forgotten and dissolution test data are frequently taken as *prima facie* evidence of the *in vivo* performance of the drug product, especially when designing modified or sustained release products. While it might be argued that for an immediate release tablet designed to disintegrate upon contact with stomach contents this might be an acceptable conclusion, for almost any other type of product such *in vivo–in vitro* correlation (IVIVC) is likely to be wildly optimistic.

The first approximation involves the volume of dissolution medium. The stomach contents in the fed state have been reported to be 686 ± 93 mL (total volume of contents, including solids), reducing to 45 ± 18 mL in the fasted state (Schiller *et al.*, 2005), and as material progresses along the GI tract its water content reduces. The same authors show that in the small intestine fluid volumes reduce to 54 ± 41 mL (fed) and 105 ± 72 mL (fasting) and in the large intestine to 11 ± 26 mL (fed) and 13 ± 12 mL (fasting). One of the reasons the fluid volumes are smaller in the fed state in the small and large intestines is that the measurement accounts only for free water, not water bound to solid matter. Moreover, MRI imaging showed that even these small amounts of fluid were not distributed homogeneously, but as small pockets. Effectively, this means that a solid drug product will be either immersed in fluid or in an essentially dry environment as it transits along the GI tract.

Secondly, McConnell, Fadda and Basit (2008) note that gastrointestinal fluids are complex, dynamic and fluctuating, in complete contrast to well-defined dissolution buffers. They suggest using buffers with ionic strengths closer to gastric fluids (for instance, Hank's and Kreb's bicarbonate buffers) as a better way of correlating dissolution data with *in vivo* performance. In addition, they advocate the inclusion of bile salts and phospholipids in the dissolution buffer. The dissolution of ionisable compounds is hugely affected by the ionic strength and ionic composition of the dissolution medium.

Finally, in a standard dissolution test the concentration of drug substance in solution increases with time to a maximum. *In vivo*, conversely, drug substance may be adsorbed, absorbed or metabolised, all processes that will act to increase the rate of dissolution in accordance with Equation (5.3).

Summary box 5.1

- Dissolution is described by the NWNB equation.

- The model assumes that dissolving molecules must diffuse across a static boundary layer to reach bulk solvent.

- The solution is saturated in the immediate vicinity of the dissolving solid.

- Dissolution rates are measured with a dissolution test.

- Standard tests are defined in pharmacopoeia.

- Dissolution tests are operated under sink conditions (the concentration of the dissolving solid never increases above 10% of the solubility).

- Dissolution tests were introduced as an assay for batch-to-batch quality assurance and are not usually representative of *in vivo* behaviour.

5.3.1   Intrinsic dissolution rate (IDR)

One assumption in the use of Equation (5.3) is that the parameters D, A and h are constant. Assuming a constant stirring speed and that the solution does not increase in viscosity as the solid dissolves, this is appropriate for D and h but A must always change as the solid dissolves (if a tablet disintegrates, for instance, then A would increase rapidly at the start of dissolution before decreasing to zero) and there will be a concomitant effect on the dissolution rate.

If the experiment is constructed such that A remains constant throughout dissolution and sink conditions are maintained, so that $(S_t - C) \cong S_t$, then the measured rate is called the *intrinsic dissolution rate* (IDR) and Equation (5.3) reduces to

$$IDR = KS_t \qquad (5.4)$$

Note here that IDR is usually determined for pure materials, not formulated products containing excipients. Construction of a dissolution test to measure IDR requires a method of sample preparation that keeps the surface area of dissolving solid constant. This is most easily achieved by allowing material to dissolve from only one face of a compact (which requires the other faces to be covered, typically by wax coating or by mounting the compact in a specially designed holder, made of metal or Teflon for instance). Viegas *et al.* (2001) discuss the use of two types of apparatus (rotating disc and stationary disc) for measuring IDR. Once mounted, dissolution data are recorded as usual.

Study question 5.3   How is IDR determined from the dissolution data?

5.3.2   IDR as a function of pH

Measurement of IDR as a function of either pH or ionic strength can give good insight into the mechanism of drug release and the performance improvement of salt forms since, for weak acids, substitution of Equation (2.12) into Equation (5.4) yields

$$IDR = K(S_0 [1 + \text{antilog}\,(\text{pH} - pK_a)]) \tag{5.5}$$

and for weak bases substitution of Equation (2.18) into Equation (5.4) yields

$$IDR = K(S_0 [1 + \text{antilog}\,(pK_a - \text{pH})]) \tag{5.6}$$

In either case the measured IDR will clearly be affected either by the pH of the medium or the microenvironment surrounding the solid surface created by the dissolving solute.

The effect of pH on IDR is easily established by selection of dissolution media. Standard media (0.1 M HCl, phosphate buffers, etc.) can be used or, in order to get a more realistic insight into dissolution rates *in vivo*, simulated or real gastrointestinal fluids can be employed.

If the drug is an acid or base, then the self-buffering effect upon dissolution should not be ignored. In particular, the saturated concentration of solute in the diffusion layer often means that the pH in the medium immediately surrounding the dissolving solid differs significantly from that of the bulk solvent and will lead to deviations from the ideal behaviour predicted by Equations (5.5) and (5.6). A schematic representation of the buffering effect of salicylic acid (Serajuddin and Jarowski, 1985) is shown diagrammatically in Figure 5.5, where it can be seen that the pH of the solution near to the surface of the dissolving solid differs considerably from that of the bulk solvent. This effect is particularly important when considering dissolution of salts and

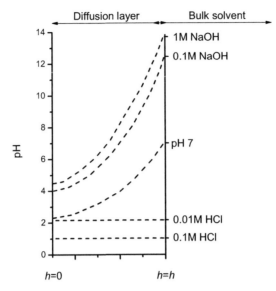

Figure 5.5 pH across the diffusion layer for dissolution of salicylic acid in various media, showing that the pH at the surface of the dissolving solid can be significantly different from that of the bulk solvent (redrawn from Serajuddin and Jarowski (1985), with permission from John Wiley & Sons, Inc.).

means that selection of the buffer system is extremely important; dissolution may well proceed at different rates in two buffers at the same pH but with different ionic strengths or with different counterions.

5.3.3   IDR and the common ion effect

The common ion effect should not be ignored, especially for hydrochloride salts, as the chloride ion is often present in reasonably high concentrations in body fluids (0.1 M in gastric fluid and 0.13 M in intestinal fluid; Lindahl *et al.*, 1997). For this reason, fed and fasted simulated intestinal fluids should contain 0.1 and 0.2 M Cl⁻ respectively (Dressman and Reppas, 2000).

The equilibrium for dissolution of a hydrochloride salt is

$$BH^+Cl^-_{(s)} \rightleftharpoons BH^+_{(aq)} + Cl^-_{(aq)} \tag{5.7}$$

As before, the equilibrium constant may be written as

$$K = \frac{[BH^+][Cl^-]}{[BH^+Cl^-]} \tag{5.8}$$

and since the concentration of the solid is effectively constant then

$$K_{sp} = [\text{BH}^+][\text{Cl}^-] \tag{5.9}$$

where K_{sp} is the solubility product. In more general terms, a solubility product expression can be written for all salts of the form $A_x^+ B_y^-$:

$$K_{sp} = [A^+]^x [B^-]^y \tag{5.10}$$

When the product of the ion concentrations exceeds the value of K_{sp} then solid salt will be precipitated. Solubility products may be written for all salts in a given system. If a system contains two salts and one of the ions is common between the salts, then dissociation of both salts will be reduced because the concentration of the common ionic species will increase in both solubility product equations.

Returning to chloride salts, when the concentration of $\text{Cl}^-_{(aq)}$ is high the equilibrium in Equation (5.7) will be shifted to the left-hand side and the solubility advantage of the salt is diminished. Li *et al.* (2005) demonstrated the effect of chloride concentration on the IDR of haloperidol salts and showed that dissolution of the hydrochloride salt was slower than that of either the phosphate or mesylate salt. Interestingly, they also showed that there was an effect of chloride concentration on the dissolution rate of the phosphate and mesylate salts (because both salts converted to the hydrochloride salt at the surface of the dissolving disc). Solubility products are important in the formation of salts, the subject of Chapter 6.

Summary box 5.2

- If the surface area of the dissolving compact of pure drug substance is kept constant, the intrinsic dissolution rate is measured.

- Where the solute is a weak acid or base dissolution will result in a change in solution pH.

- The pH of the boundary layer may be different from that of the bulk solvent and will ultimately control the observed dissolution rate.

- Care must be taken to ensure that the ionic strength of the dissolution medium does not affect the dissolution rate and that the common ion effect is considered.

5.4 Summary

The standard model of dissolution assumes that molecules leaving the surface of the dissolving solid must diffuse through a stationary boundary layer to reach the bulk solvent. Where a solute is acidic or basic in nature there will be a buffering effect upon dissolution, which means that the local pH of the solution surrounding the solid surface is not equivalent to that of the bulk solvent. The common ion effect, especially that of the chloride ion, should not be ignored during dissolution under physiological conditions. The interpretation of dissolution test data as being reflective of *in vivo* behaviour is extremely dangerous without knowledge of the behaviour of the drug substance under physiological conditions.

References

Brunner, E. (1904) Reaktionsgeschwindigkeit in heterogenen systemen. *Z. Phys. Chem.*, **43**, 56–102.

Brunner, E. and Tolloczko, S. (1900) Über die Auflösungsgeschwindigkeit fester Körper. *Z. Phys. Chem.*, **35**, 283–290.

Danckwerts, P.V. (1951) Significance of liquid-film coefficients in gas absorption. *Ind. Engng Chem.*, **43**, 1460–1467.

Dokoumetzidis, A. and Macheras, P. (2006) A century of dissolution research: from Noyes and Whitney to the biopharmaceutical classification system. *Int. J. Pharm.*, **321**, 1–11.

Dressman, J.B. and Reppas, C. (2000) *In vitro–in vivo* correlations for lipophilic, poorly water-soluble drugs. *Eur. J. Pharm. Sci.*, **11**, S73–S80.

Klein, S. (2006) The mini paddle apparatus – a useful tool in the early developmental stage? Experiences with immediate-release dosage forms. *Dissol. Technol.*, **13**, 6–11.

Li, S., Doyle, P., Metz, S. *et al.* (2005) Effect of chloride ion on dissolution of different salt forms of haloperidol, a model basic drug. *J. Pharm. Sci.*, **94**, 2224–2231.

Lindahl, A., Ungell, A.L. and Lennernas, H. (1997) Characterization of fluids from the stomach and proximal jejunum in men and women. *Pharm. Res.*, **14**, 497–502.

McConnell, E.L. Fadda, H.M. and Basit, A.W. (2008) Gut instincts: explorations in intestinal physiology and drug delivery. *Int. J. Pharm.*, **364**, 213–226.

Nernst, W. (1904) Theorie der Reaktionsgeschwindigkeit in Heterogenen Systemen. *Z. Phys. Chem.*, **47**, 52–55.

Noyes, A.A. and Whitney, W.R. (1897) The rate of solution of solid substances in their own solutions. *J. Am. Chem. Soc.*, **19**, 930–934.

Schiller, C., Fröhlich, C.-P., Giessmann, T. *et al.* (2005) Intestinal fluid volumes and transit of dosage forms as assessed by magnetic resonance imaging. *Ailment Pharmacol. Theory*, **22**, 971–979.

Serajuddin, A.T.M. and Jarowski, C.I. (1985) Effect of diffusion layer pH and solubility on the dissolution rate of pharmaceutical acids and their sodium salts. II: salicylic acid, theophylline and benzoic acid. *J. Pharm. Sci.*, **74**, 148–154.

Viegas, T.X., Curatella, R.U., Van Winkle, L.L. and Brinker, G. (2001) Measurement of intrinsic drug dissolution using two types of apparatus. *Pharm Technol.*, **25**, 44–43.

Wilderman, M. (1909) Über die Geschwindigkeit Molekularer und Chemischer Reaktionen in Heterogenen Systemen. *Erster Teil. Z. Phys. Chem.*, **66**, 445–495.

Answers to study questions

5.1 Removing aliquots of solution for analysis will reduce the total volume of dissolution medium in which the drug product can dissolve. Good practice would be to replace each aliquot with an equal volume of fresh dissolution medium, but it should be recognised that this will dilute the concentration (slightly) and increase the dissolution rate. Applying a correction factor would obviate this effect, as would in-line or in situ measurement of concentration.

5.2 Possibly, but there are other variables in the NWNB equation that will affect the dissolution rate, principally the change in surface area, which is likely to decrease with time. If the surface area of the dissolving solid is kept constant, then intrinsic dissolution rate is measured.

5.3 The gradient of the dissolution line divided by the surface area of the compact gives the IDR.

6
Salt Selection

6.1 Introduction

If a drug substance has poor aqueous solubility but is a weak acid or base, then conversion to a salt form may offer a convenient method for solubility improvement (although other factors may also drive the desire for salt formation, including as a processing step to allow isolation of the active, to improve stability and/or to aid purification). A number of other physicochemical properties will change upon formation of a salt (Table 6.1). Berge, Bighley and Monkhouse (1977) and Serajuddin (2007) provide excellent reviews of the physicochemical properties of pharmaceutical salts and the *Handbook of Pharmaceutical Salts* (Stahl and Wermuth, 2011) is an invaluable resource for preformulation scientists considering preparing a salt form.

Not all the physicochemical changes that arise as a result of salt formation may be beneficial and so a decision must be made early during preformulation as to which salt form (if any) is to be taken into development and the decision will not reduce to an absolute improvement in aqueous solubility alone. In particular, salt selection should be made before commencement of toxicity testing, because of the associated cost and potential time delay in development of switching to a different salt form. It must also be considered that the regulatory authorities may consider a new salt form as changing the chemical nature of the active moiety, so it may need to be filed as a new chemical entity.

The prevalence of salt forms of drug substances (estimated at ca. 50%; Kumar, Amin and Bansal, 2008) suggests, however, that the benefits usually outweigh the drawbacks. This chapter will focus on the principles of salt formation, salt selection strategy, salt screening and the fundamental reasons for improved solubility and dissolution. Characterisation of physical forms will be considered in later chapters.

Essentials of Pharmaceutical Preformulation, First Edition. Simon Gaisford and Mark Saunders.
© 2013 John Wiley & Sons, Ltd. Published 2013 by John Wiley & Sons, Ltd.

Table 6.1 Possible advantages and disadvantages of salt formation.

Advantages	Disadvantages
Enhanced solubility	Decreased percentage of active
Increased dissolution rate	Increased hygroscopicity
Higher melting point	Decreased chemical stability
Lower hygroscopicity	Increased number of polymorphs
Improved photostability	Reduced dissolution in gastric media
Better taste	No change in solubility in buffers
Higher bioavailability	Corrosiveness
Better processability	Possible disproportionation
Easier synthesis or purification	Additional manufacturing step
Potential for controlled release	Increased toxicity

6.2 Salt formation

A salt is formed when an acid reacts with a base, resulting in a species held together by ionic bonds. In principle any weak acid or base can form a salt, although in practice if the pK_a of the base is very low the salt formed is unlikely to be stable at physiological pH. In a review of the salt forms of 203 current basic drugs Stephenson, Aburub and Woods (2011) note that no marketed salt exists for a basic drug substance with a pK_a below 4.6, for this reason, and suggest that 5 is a general value below which salt formation is unlikely to be effective.

Because they usually dissociate rapidly upon dissolution into water, salts are considered electrolytes, although in some instances a drug substance sounds like a salt but is in fact a single entity bound via covalent bonds (fluticasone propionate for instance) in which case electrolytic behaviour does not apply. Many drug substances are available in different salt forms (diclofenac is marketed in both sodium and potassium salt forms for instance) and yet they are marketed as being therapeutically indistinguishable, which may not be a valid assumption (Patel *et al.*, 2009). Table 6.2 shows some physicochemical data for five salt forms of diclofenac, which illustrate this point.

Acids and bases can be classified as strong through to extremely weak, based on their pK_a (Table 6.3). When strong acids react with strong bases the

Table 6.2 Physicochemical properties of some diclofenac salts (data from Fini *et al.* (1996)).

Salt	MW	Melting point (°C)	Solubility (mg mL^{-1})	% Dissolution after 2 h
Li	303	>300	8.2	97
Na	318	283–285	9.6	87
K	334	296–298	4.7	82
Rb	381	>300	7.8	80
Cs	428	275–277	6.2	68

Table 6.3 Descriptions of acid and base strength (data from Stahl and Wermuth (2011)).

	pK_a	
Description	Acid	Base
Very strong	<0	>14
Strong	0–4.5	9.5–14
Weak	4.5–9.5	4.5–9.5
Very weak	9.5–14	0–4.5
Extremely weak	>14	<0

reaction tends to completion, as both species will be fully ionised, a process termed *neutralisation*. For example,

$$HCl + NaOH \rightarrow NaCl + H_2O \tag{6.1}$$

In this instance the salt formed will precipitate, once it is present at a concentration beyond its solubility at any particular pH. Most drug substances, however, are either weak acids or bases, in which case acidic and basic character is usually based on the Brønsted–Lowry definition. This posits that an acidic compound is a proton donor and a basic compound is a proton acceptor. The removal of a proton from an acid produces a *conjugate base* (A^-) and addition of a proton to an acceptor produces a *conjugate acid* (BH^+):

$$HA + H_2O \rightleftharpoons H_3O^+ + A^- \tag{6.2}$$
$$B + H_3O^+ \rightleftharpoons BH^+ + H_2O \tag{6.3}$$

Note that the Brønsted–Lowry definition requires acidic species to have an ionisable proton but does not require basic compounds to possess a hydroxide group, simply that they can accept a proton (the theory does not consider KOH to be a base, for instance, but a salt containing the basic OH^- moiety). Note also that protons do not exist in aqueous solution in isolation but as the hydronium ion. In the case of a weak base (B) reacting with a strong acid, the conjugate acid and conjugate base may then form a salt:

$$B + HCl \rightleftharpoons BH^+ + Cl^- \rightleftharpoons BH^+Cl^- \tag{6.4}$$

When a salt dissolves in water it will dissociate (in the majority of cases, with exceptions including some mercury salts). Assuming dissolution of a basic salt then the species in solution is the conjugate acid. The conjugate acid can donate its proton to water, reforming the free base:

$$BH^+ + H_2O \rightleftharpoons B + H_3O^+ \tag{6.5}$$

All of the reasons for the change in solubility of salts are encompassed in Equation (6.5). A basic salt contains the conjugate acid of the drug substance. Upon dissolution the conjugate acid donates its proton to water and the free base is formed. The solute is thus the free base, but the pH of the solution in which it is dissolved has reduced because of the donated proton. Recall from Chapter 4 that the solubility of weak bases increases as the pH of the solution reduces. Thus dissolution of a basic salt increases solubility over the free form because there is a concomitant reduction in pH of the solution.

Since the pH of a solution of a dissolved acid is given by

$$pH = \frac{1}{2}(pK_a - \log[\text{acid}]) \tag{6.6}$$

It follows that because the acid species is BH^+ then

$$pH = \frac{1}{2}(pK_a - \log[BH^+]) \tag{6.7}$$

Example 6.1   What is the pH of a 0.2M solution of ergotamine tartrate (pK_a 6.25)?

From Equation (6.7),

$$pH = \frac{1}{2}[(6.25) - \log(0.2)] = 3.48$$

Figure 6.1 shows the solution pH as a function of concentration for basic salts over the pK_a range 9–6, calculated with Equation (6.7). It is apparent that even a small concentration of salt leads to a significant change in solution pH.

A similar situation occurs for the reaction of a weak acid with a strong base:

$$HA + NaOH \rightleftharpoons NaA + H_2O \tag{6.8}$$

Upon dissolution of an acidic salt the conjugate base is formed:

$$NaA \rightleftharpoons Na^+ + A^- \tag{6.9}$$

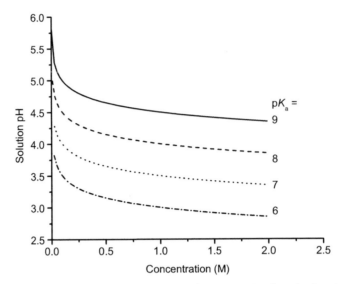

Figure 6.1   Solution pH as a function of concentration for a basic salt.

The conjugate base can then accept a proton from a hydronium ion, reforming the free acid and increasing the pH of the solution:

$$A^- + H_3O^+ \rightleftharpoons AH + H_2O \tag{6.10}$$

The solute is thus the free acid, but the pH of the solution in which it is dissolved has increased because a hydronium ion has been removed from the system. Recall again from Chapter 4 that the solubility of weak acids increases as the pH of the solution increases.

The pH of a solution of base is given by

$$pH = \frac{1}{2}(pK_a + pK_w + \log[\text{base}]) \tag{6.11}$$

Since A^- is a conjugate base then

$$pH = \frac{1}{2}(pK_a + pK_w + \log[A^-]) \tag{6.12}$$

where pK_w is the self-ionisation constant of water, usually assigned a value of 14, although the value shows considerable temperature dependence (Table 6.4). This is why neutral pH is 7.0 only at 24 °C and ranges from 7.5 at the freezing temperature of water to a little over 6.0 at the boiling temperature of water. Figure 6.2 shows the solution pH as a function of concentration for acidic salts over the pK_a range 3–6, calculated with Equation (6.12).

Table 6.4 Values for the self-ionisation constant of water as a function of temperature (data from Bandura and Lvov (2005)).

Temperature (°C)	0	25	50	75	100
K_w	14.95	13.99	13.26	12.70	12.25

Example 6.2 What is the pH of a 0.1 M solution of diclofenac sodium (pK$_a$ 4.0)?

From Equation (6.12),

$$pH = \frac{1}{2}(4.0 + 14 + \log[0.1]) = 8.5$$

Several consequences arise from this discussion. One is that salt formation might not be best achieved in an aqueous solution, since dissolution of a salt in water generally results in formation of the free acid or base. For this reason salts are often formed in organic solvents (although it has been suggested that the addition of at least 10% v/v water to crystallisation solvents can improve salt formation for poorly water soluble bases; Tarsa *et al.*, 2010). Secondly, the increase in solubility of a salt over the corresponding free acid or base is a result only of the change in pH upon dissolution. *The intrinsic solubility of the free acid or base does not change.*

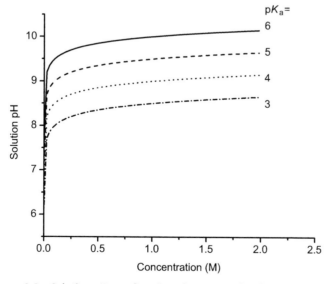

Figure 6.2 Solution pH as a function of concentration for an acidic salt.

Study question 6.1   What does this mean for solubility of salts in buffered media?

Summary box 6.1

- Salts form when acids react with bases. Weak acids are reacted with strong bases and vice versa.

- Acidic drugs are proton donors, basic drugs are proton acceptors.

- Forming a salt does not change the pK_a of the parent molecule.

- The increase in solubility of a salt is because of the change in pH upon dissolution. Dissolution of an acidic salt *increases* pH and dissolution of a basic salt *decreases* pH.

6.2.1   Selection of a salt-forming acid or base

In order for the salt to form there must be a sufficient difference in pK_a values between the acid and base (the *reactivity potential*). For the transfer of a proton from an acid to a weak base the pK_a of the acid must be less than that of the weak base and vice versa. As a general rule, a difference in pK_a (ΔpK_a) of 3 is indicated, although salt formation can occur with smaller differences (for instance, doxylamine succinate forms spontaneously even though ΔpK_a is only 0.2; Wells, 1988[1]). The reason for this is to ensure that both species are ionised in solution, thus increasing the chance of interaction. Childs, Stahly and Park (2007) note that if ΔpK_a is greater than 3 then salts will invariably form while if ΔpK_a lies between 3 and 0 then knowledge of ΔpK_a per se is not predictive of whether salt formation will occur. In cases where ΔpK_a is less than zero co-crystal (Chapter 7) formation is the more likely outcome (Bhogala, Sasavoju and Nangia, 2005).

Brittain (2007, 2008) points out that if one considers the formation of the salt of a weak acid (such as ibuprofen) with a strong base,

$$H - Ibu + B \rightleftharpoons HB^+ + Ibu^-$$ (6.13)

Then the position of equilibrium is given by

$$K_s = \frac{[Ibu^-][HB^+]}{[H - Ibu][B]}$$ (6.14)

---

[1] Since the difference in pK_a values reflects the difference in free energy, the fact that this salt forms spontaneously implies that there is either a large entropic or a large enthalpic thermodynamic driving force.

and so

$$K_s = \frac{K_a K_b}{K_w} \qquad (6.15)$$

or

$$pK_s = pK_a + pK_b - pK_w \qquad (6.16)$$

Expressed in terms of pK_a only, Equation (6.16) becomes

$$pK_s = pK_{a,acid} - pK_{a,base} \qquad (6.17)$$

Thus, with knowledge of the pK_a values of the drug substance and salt-forming acid or base (as appropriate) it is possible to calculate the position of equilibrium of the salt formation reaction. Assuming that equimolar concentrations of acid and base are reacted initially (C) and that upon attainment of equilibrium some fraction of salt (x) is formed, then Equation (6.14) becomes

$$K_s = \frac{x^2}{(C-x)^2} \qquad (6.18)$$

If C is set equal to 1, so fractions are calculated, rearranging Equation (6.18) leads to the quadratic form:

$$(K_s - 1)x^2 - 2K_s x + K_s = 0 \qquad (6.19)$$

which can be solved for x:

$$x = \frac{2K_s - \sqrt{(2K_s)^2 - 4[(K_s - 1)K_s]}}{2(K_s - 1)} \qquad (6.20)$$

Equation (6.20) allows determination of the fraction of salt formed as a function of ΔpK_a between the acid and base. The data in Figure 6.3, generated with Equation (6.20), show the fraction of salt formed (as a percentage) versus the acid pK_a for a basic drug substance of pK_a 7.5. The general rule noted above that ΔpK_a should be greater than 3 is explained by this relationship, because when $\Delta pK_a = 3$ nearly 97% of the reactants will form a salt (and if $\Delta pK_a > 4$ more than 99% of the reactants form a salt). Brittain (2008) shows that the same relationship is true for an acidic drug. Thus, selection of a salt former starts with knowledge of its pK_a and the pK_a of the drug substance. The pK_a values of some of the most common salt-forming acids and bases are given in Tables 6.5 and 6.6 (note that these values are determined in water and thus will change if organic solvents are used, which is likely). The top 10 anions and cations by frequency for drugs in the 2006 USP are shown in Table 6.7.

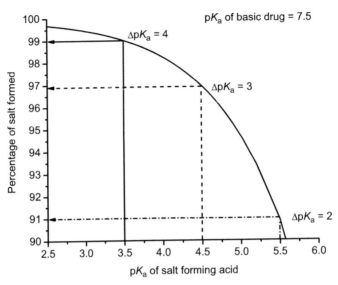

Figure 6.3 Percentage of salt formed as a function of acid pK_a when reacted with a basic drug of pK_a 7.5.

Table 6.5 Values of pK_a for selected pharmaceutical acids (data from Stahl and Wermuth (2011)).

Acid	Anion	pK_a	Example
Hydrobromic	Hydrobromide	<−6.0	Galantamine
Hydrochloric	Hydrochloride	−6.0	Clindamycin
Sulphuric	Sulphate	−3.0, 1.92	Salbutamol
p-Toluenesulphonic	Tosylate	−1.34	Sorafenib
Nitric	Nitrate[a]	−1.32	Miconazole
Methanesulphonic	Mesylate	−1.2	Benztropine
Naphthalene-2-sulphonic	Napsylate	0.17	Levopropoxyphene
Benzenesulphonic	Besylate	0.7	Amlodipine
Oxalic	Oxalate	1.27, 4.27	Escitalopram
Maleic	Maleate	1.92	Fluvoxamine
Phosphoric	Phosphate	1.96, 7.12, 12.32	Fludarabine
Pamoic	Pamoate	2.51, 3.1	Amitriptyline
Tartaric	Tartrate	3.02, 4.36	Metoprolol
Fumaric	Fumarate	3.03, 4.38	Formoterol
Citric	Citrate	3.13, 4,76, 6.40	Sildenafil
Hippuric	Hippurate	3.55	Methenamine
Benzoic	Benzoate	4.19	Emamectin
Succinic	Succinate	4.21, 5.64	Metoprolol
Acetic	Acetate	4.76	Megestrol
Carbonic	Carbonate	6.46, 10.3	Lithium

[a]No longer considered acceptable for pharmaceutical use.

Table 6.6 Values of pK_a for selected pharmaceutical bases (data from Stahl and Wermuth (2011)).

Base	Cation	pK_a	Example
Potassium hydroxide	Potassium	ca. 14	Benzylpenicillin
Sodium hydroxide	Sodium	ca. 14	Diclofenac
Zinc hydroxide	Zinc	ca. 14	Bacitracine
Calcium hydroxide	Calcium	12.6, 11.57	Fenoprofen
Magnesium hydroxide	Magnesium	11.4	Menbutone
Choline	Choline	>11	Theophylline
Lysine	Lysine	10.79, 9.18, 2.16	Ibuprofen
Benzathine	Benzathine	9.99, 9.39	Ampicillin
Piperazine	Piperazine	9.82, 5.58	Naproxen
Meglumine	Meglumine	9.5	Flunixin
Ammonia	Ammonium	9.27	Glycyrrhizinic acid
Tromethamine	Trometamol	8.02	Lodoxamide
Aluminium hydroxide	Aluminium	>7	

[a]No longer considered acceptable for pharmaceutical use.

Study question 6.2 Why do you think the hydrochloride salt is the most common form for basic drug substances? What disadvantages might using this salt have?

Stahl and Wermuth (2011) organise salt formers into three categories, which may be used as a guide to selection:

- *First class* salt formers are those that form physiologically ubiquitous ions or that occur as metabolites in biochemical pathways (including the hydrochloride and sodium salts). As such, they are considered to be unrestricted in their use.

Table 6.7 Frequency of pharmaceutical anions and cations of drugs in USP 29-NF24 (data from Kumar *et al.* (2008)).

Anion	Frequency (%)	Cation	Frequency (%)
Hydrochloride	39.96	Sodium	62.79
Sulphate	10.58	Potassium	11.05
Acetate	6.70	Calcium	8.72
Phosphate	4.97	Aluminium	4.65
Chloride	4.54	Benzathine	2.33
Maleate	3.67	Meglumine	2.33
Citrate	3.02	Zinc	2.33
Mesylate	2.59	Magnesium	1.74
Succinate	2.38	Tromethamine	1.74
Nitrate[a]	2.38	Lysine	1.16

- *Second class* salt formers are those that are not naturally occurring but that have found common application and have not shown significant toxicological or tolerability issues (such as the sulphonic acids).

- *Third class* salt formers are those that are used in special circumstances to solve a particular problem. They are not naturally occurring nor in common use.

An additional factor to consider is that the salt formed should exist as a crystalline solid, to enable ease of isolation and purification. Amorphous salts are highly likely to cause problems in development and use and so should be avoided.

6.2.2   Salt screening

Once potential salt formers have been selected they must be combined with the drug substance in order to see which preferentially form salts. Since the potential number of permutations and combinations of salt formers and solvents is large, a convenient method for salt screening is to use a microwell plate approach. A small amount of drug substance (ca. 0.5 mg) in solvent is dispensed into each well of a 96-well plate. To each well is added a solution of potential counterion. It is possible to construct the experiment so the effect of the solvent is examined in the x dimension and the effect of the counterion is examined in the y dimension. Solvents should be selected carefully and those listed as class I by ICH Guideline Q3C(R5) (2011) (i.e. those that are either known or suspected carcinogens or that pose an environmental danger) should be avoided. Typically used solvents are listed in Table 6.8, with their ICH class.

After an appropriate length of time, the presence in each well of salt crystals is checked with an optical device (for instance, a microscope or a nephelometer). If no crystals are seen then the plate can be stored at a lower temperature. If the reduction in temperature does not cause precipitation then as a last attempt the temperature can be increased to evaporate the solvent (although care must be taken in this case during subsequent analysis because the isolate may contain a simple mixture of unreacted drug and salt former, rather than the salt itself).

Once a potential salt has been identified preparation can be undertaken with slightly larger sample masses (10–50 mg). X-ray powder diffraction (XRPD) may be used to get a preliminary idea of polymorphic form while melting points may be determined with a melting point apparatus, hot-stage microscopy (HSM) or differential scanning calorimetry (DSC). Examination with HSM, if operated under cross-polarised filters, allows quick visual confirmation of crystallinity and melting and any other changes in physical form

Table 6.8 Properties of some common solvents used for salt screening (data from Huang and Tong (2004)).

Solvent	Boiling point (°C)	Dielectric constant (ε)	ICH class
N,N-Dimethylformamide	153	37.0	2
Acetic acid	118	6.2	3
Water	100	78.4	n/a
1-Propanol	97	20.3	3
2-Propanol	83	19.9	3
Acetonitrile	82	37.5	3
2-Butanone	80	18.5	3
Ethanol	78	24.6	3
Ethyl acetate	77	6.0	3
n-Hexane	69	1.9	2
Isopropyl ether	68	3.9	4
Methanol	65	32.2	2
Acetone	57	20.7	3
Methylene chloride	40	8.9	2
Diethyl ether	35	4.3	3

during heating, while analysis with DSC provides the enthalpy of fusion in addition to the melting temperature (and so access to ideal solubility). Additional analyses with thermogravimetry analysis (TGA) and dynamic vapour sorption (DVS) will provide information on residual water content, the presence of the salt in a solvated or hydrated state and/or hygroscopicity tendency. All of these experiments can be performed with just ca. 50 mg of salt, if available.

Of course a salt screen such as this may result in a large number of potential salt candidates and so an approach is needed to select a drug substance for development. Morris et al. (1994) published an integrated approach to salt candidate selection which is effectively a decision tree method based on easily measured physicochemical properties (Figure 6.4). Following isolation of crystalline salts the hygroscopicity of each is determined (since this is easy to measure, does not in principle destroy the sample and gives a likely indication of stability and processability). Those salts with acceptable properties are advanced to tier 2, where physical form stability with respect to relative humidity (RH) and solubility are measured. This would indicate the likelihood that the salt would change polymorphic form during processing or storage and allows selection of a salt with satisfactory solubility (bearing in mind that the salt with the highest solubility would not necessarily be the easiest to manufacture or formulate). Finally tier 3 testing involves thermal and photostability assessment under accelerated stress conditions. At each tier the time and cost associated with testing increases, so it is good practice to reduce the number of candidates at each stage.

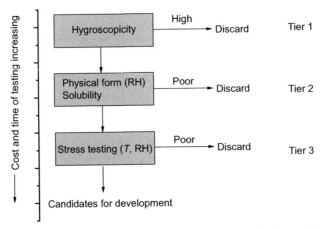

Figure 6.4 A decision tree for salt selection based on characterisation of physicochemical properties (data from Morris *et al.* (1994)).

Of course, such a regimented scheme must be undertaken with an open mind. The calcium and sodium salts of fenoprofen have equivalent bioavailability, distribution and elimination profiles, but the calcium salt is less hygroscopic so it is the marketed form (Lemke *et al.*, 2008). Ranitidine hydrochloride, conversely, is extremely hygroscopic (Teraoka, Otsuka and Matsuda, 1993) yet is a successful product. Similarly, sertraline hydrochloride has around 28 reported polymorphic forms (Remenar *et al.*, 2003). Both of the latter drugs would have been discarded according to the scheme in Figure 6.4.

Summary box 6.2

• Salt formation can be used as a method of purification of the drug.

• For salt formation there should be a difference in pK_a between acid and base of at least 2, while 3 is preferable.

• Salts that form physiological ions are preferable but solubility can be reduced because of the common ion effect.

• Salts are often formed in nonaqueous solvents, since they must precipitate in order to be isolated. Cannot use ICH class I solvents.

6.3 Salt solubility

It is not a simple matter to predict the solubility of a salt form, for many of the reasons outlined in earlier chapters; in particular, the common ion effect

Table 6.9 Solubilities of various procaine salts as a function of temperature (data from Guerrieri *et al.* (2010)).

Procaine salt	Solubility at 25 °C (mol L^{-1})	Solubility at 50 °C (mol L^{-1})
Phosphate	0.488	0.87
Citrate	0.697	1.2
Besylate	0.901	2.14
Oxalate	0.067	0.238
Bisulphate	2.64	3.55
Esylate	2.33	3.0
Napsylate	0.026	0.049
Tosylate	0.068	0.358
Formate	2.89	3.75
Mesylate	2.78	3.29
HCl	2.58	3.84

cannot be ignored, especially when dissolution and solubility in biological fluids are considered. Table 6.9 shows the aqueous solubilities of a number of procaine salts; no discernible trend in solubility with cation is seen. There are many empirical approaches in the literature for estimating the solubility of salts, but most require knowledge of the melting point of the salt, a value most reliably determined by preparing the salt and melting it (in which case, the salt is available for solubility determination by experiment). This section will thus consider the underlying principle of solubility pH dependence based on ionic equilibria and assumes that solubility would be determined experimentally using the actual salt.

6.3.1 Solubility of basic salts

Dealing with a basic salt first, at high pH the solubility will be equal to that of the un-ionised (or free) base (i.e. at its lowest) and at low pH the solubility will be equal to that of the ionised base (i.e. at its highest). There will be a region between these extremes where the solubility will vary with pH (Figure 6.5). The standard interpretation of a solubility profile of this form is based on the model of Kramer and Flynn (1972), who assumed that the overall profile is the sum of two solubility profiles. In region 1, the dissolved solute is in equilibrium with solid salt and in region 2 the dissolved solute is in equilibrium with the solid free base. The point at which the two solubility profiles intersect is termed pH$_{max}$.

The solubility profile in region 2 is analogous to that discussed in Chapter 2 for weak bases, being the sum of the concentration of saturated free base [B]$_s$ and the ionised base, and so is given by

$$S_{t,base,pH>pH_{max}} = [B]_s \left[1 + antilog(pK_a - pH)\right] \tag{6.21}$$

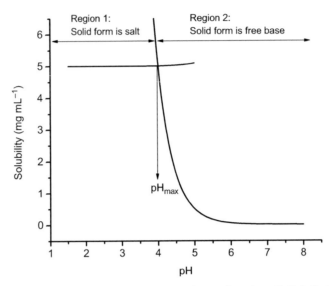

Figure 6.5   Solubility profile for a basic salt as a function of pH (pK_a 6.7).

In region 1 a similar situation exists, now between the saturated conjugate acid $[BH^+]_s$ and ionised base, which has the effect of reversing the terms in Equation (6.21):

$$S_{t,base,pH<pH_{max}} = [BH^+]_s \left[1 + \text{antilog}(pH - pK_a)\right] \qquad (6.22)$$

The data shown in Figure 6.5 were generated using Equation (6.21) (below pH_{max}) and Equation (6.22) (above pH_{max}) and assume a pK_a of 6.7 and an intrinsic solubility of 0.01 mg mL^{-1}.

The data suggest that a basic salt will be most soluble in low pH media (such as stomach acid) but will become increasingly insoluble as pH increases (as it would in gastrointestinal fluids). Thus, if a basic salt is administered orally, its solubility will naturally decrease as it progresses along the GI tract and it will probably precipitate following gastric emptying. The bioavailability of basic drugs has also been shown to be dependent upon gastric pH. For instance, dipyridamole is less bioavailable in achlorhydric elderly patients (Russell *et al.*, 1994) while cinnarizine has reduced bioavailability in dogs with elevated gastric pH (Yamada *et al.*, 1990).

6.3.2   Solubility of acidic salts

A similar series of equations can be derived for salts of weak acids. In this case the free acid is the solid phase in equilibrium with the saturated solution

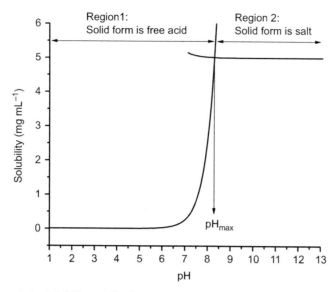

Figure 6.6   Solubility profile for an acidic salt as a function of pH (pK_a 5.6).

below pH_{max} and the salt is the solid phase in equilibrium with the saturated solution above pH_{max}. Thus the relevant equations are

$$S_{t,acid,pH<pH_{max}} = [AH]_s \left[1 + antilog(pH - pK_a)\right] \qquad (6.23)$$

$$S_{t,acid,pH>pH_{max}} = [A^-]_s \left[1 + antilog(pK_a - pH)\right] \qquad (6.24)$$

The solubility profile for an acidic salt is shown in Figure 6.6 and is opposite to the case of basic salts described above. The profile suggests that an acidic salt will be least soluble in low pH media but will become increasingly soluble as pH increases. Thus, if an acidic salt is administered orally its solubility will naturally increase as it progresses along the GI tract (indeed, its solubility in stomach acid may be so low that it will naturally dissolve only in the intestinal lumen, which can be a formulation advantage). Again, however, bioavailability will be dependent upon the gastric emptying time, since little or no absorption of acidic drugs will occur in the stomach. For instance, the rate of absorption of paracetamol is directly related to the rate of gastric emptying (Heading *et al.*, 1973). Unless absorption is particularly slow, then the rate of gastric emptying may well be the rate-limiting factor (and there may be reduced bioavailability in patients with abnormal gastric emptying).

In cases where absorption from the stomach does occur, administration with food or antacids may decrease bioavailability of acidic drugs, as the degree of ionisation will increase as the pH rises. Similarly, changes in

bioavailability may be seen in patients taking long-term H_2-antagonist ther-
apy (such as ranitidine or famotidine), because of the increase in gastric pH.

Another factor to consider is that some drugs (such as atropine or the
tricyclic antidepressants) influence the rate of intestinal motility and so may
affect the rate of absorption of many co-administered drugs. Similarly, co-
administration of a drug that induces gastric emptying (such as metoclo-
pramide) may increase absorption rates.

6.3.3   The importance of pH_{max}

At pH_{max}, which in principle is a single point on the plot, both the free
acid/base and the salt coexist in the solid phase. If the pH of a saturated
solution containing excess solid free base is lowered below pH_{max} then the
solid will convert to the salt (although the pH will not drop below pH_{max} until
enough acid has been added to convert all free base to the salt). Conversely,
if the pH of a saturated solution containing excess solid salt is raised above
pH_{max} then the solid phase will convert to the free base. The opposite holds
true for an acidic salt.

It should be apparent that pH_{max} is an important parameter and its value
will change depending upon the solubility of the salt form made. Figure 6.7
shows the solubility profiles for three salt forms of a basic drug substance and
the change in pH_{max}. For basic drug substances Equations (6.21) and (6.22)

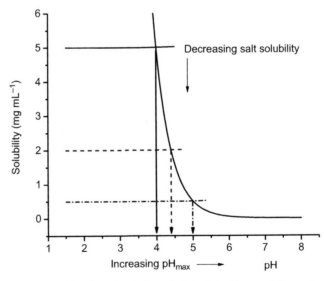

Figure 6.7 Solubility profiles for a basic drug (pK_a 6.7) in three salt forms (solubilities 5, 2 and 0.5 mg mL^{-1}).

can be set equal at pH_{max}. Solving the resulting equation leads to (Bogardus and Blackwood, 1979)

$$pH_{max} = pK_a + \log\frac{[B]_s}{\sqrt{K_{sp}}} \tag{6.25}$$

where K_{sp} is the solubility product of the salt. Equation (6.25) shows how varying pK_a, $[B]_s$ or K_{sp} will affect pH_{max} (Pudipeddi *et al.*, 2011):

- Increasing pK_a by 1 unit (making the base stronger) will increase pH_{max} by 1 unit (Figure 6.8).

- Increasing the solubility of the free base by an order of magnitude will increase pH_{max} by 1 unit (Figure 6.9).

- Increasing the solubility of the salt by an order of magnitude will decrease pH_{max} by 1 unit (Figure 6.10).

From the perspective of the species in the system, the following series of equilibria coexist at pH_{max} (for a basic hydrochloride salt):

$$B_{solid} \rightleftharpoons B + H_3O^+ + Cl^- \rightleftharpoons BH^+ + Cl^- + H_2O \rightleftharpoons BH^+Cl^-_{solid} \tag{6.26}$$

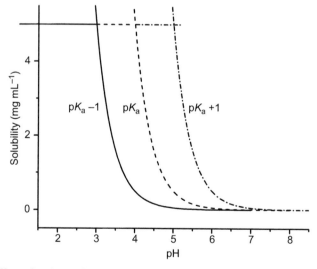

Figure 6.8 Effect of a change in pK_a on the solubility profile of a basic salt (original pK_a 6.7).

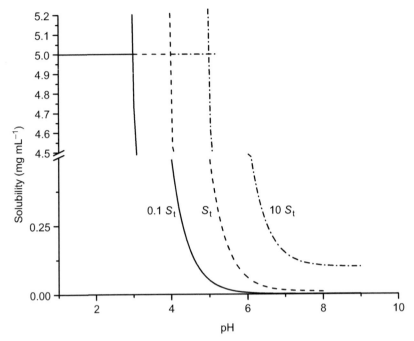

Figure 6.9   Effect of free base solubility on the solubility profile of a basic salt (S_t 0.01 mg mL^{-1}).

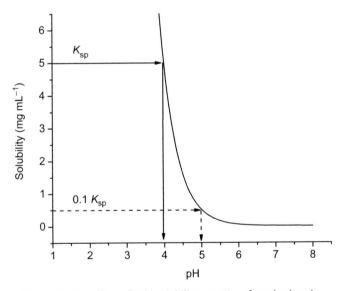

Figure 6.10   Effect of salt solubility on pH$_{max}$ for a basic salt.

If a small amount of $[H^+]$ is added to the system at pH_{max} then the equilibria shift to the right-hand side of Equation (6.26) and free base is converted to salt. Conversely, if alkali is added the equilibria shift to the left-hand side and salt is converted to free base. As the system is effectively acting as a buffer, the pH (and consequently the solubility) will not change until sufficient acid or alkali has been added to convert one solid phase completely to the other, as noted in the above discussion.

A similar analysis can be performed for an acidic salt. The value of pH_{max} can have a critical influence on the dissolution rate of salts, because the pH of the dissolution medium can cause conversion of a salt back to the free acid or base form.

Summary box 6.3

- The solubility profile of a salt has two distinct regions, one where the salt is the solid phase and the other where the free acid or base is the solid phase.

- The point of intersection of the two regions is termed pH_{max}.

- At pH_{max} the free acid or base can coexist with the salt in the solid phase. Addition of a small amount of acid or base will not change the pH of the system.

- When $pH \neq pH_{max}$ conversion between salt and free acid or base can occur.

6.4   Dissolution of salts

Dissolution rates are quantitatively described by the NWNB equation (Chapter 5):

$$\frac{dC}{dt} = \frac{DA}{Vh}(S_t - C) \tag{6.27}$$

If the experiment is performed under sink conditions, and assuming the surface area of the dosage form is kept constant, then the dissolution rate is proportional to solubility (note here that the model takes no account of wettability – dissolution of a highly soluble drug substance may be rate-limited by poor wettability and be measured to be slower than dissolution of a poorly soluble drug substance with good wettability). The NWNB model assumes formation of a saturated solution at the surface of the dissolving solid (the boundary layer) through which additional dissolving molecules must diffuse to reach the bulk solvent. Salts have the potential to increase the dissolution rate because the saturated concentration in the boundary layer is much higher than that of the free acid or base (in other words, the value of S_t is increased in Equation (6.27)). This is a direct result of the change in pH upon dissolution discussed earlier.

Since solubility is pH-dependent, the NWNB model predicts that the dissolution rate must therefore also be pH-dependent, with the solubility of the solute at the pH and ionic strength of the dissolution medium being the rate-controlling parameter. By the same argument, when the pH of the dissolution medium is around that of pH_{max} the dissolution rates of the free acid or base and its salt should be the same (because their solubilities are roughly equal at this point). There are, however, numerous examples in the literature where this is found not to be the case – examples including doxycycline hydrochloride and doxycycline (Bogardus and Blackwood, 1979), sodium salicylate and salicylic acid (Serajuddin and Jarowski, 1985) and haloperidol mesylate and haloperidol (Li *et al.*, 2005b).

These differences suggest that the pH of the solution into which the solid is dissolving (i.e. the boundary layer) is materially different from that of the bulk solvent (and so the solubility of the dissolving species is different from that expected in the bulk solvent). The difference in pH between the boundary layer and bulk solvent arises because the boundary layer is a saturated solution and because dissolution of acids, bases or salts will result in a change in pH; when saturated, the pH change is maximised. Nelson (1957) first noted this correlation during a study of the dissolution of various theophylline salts; salts with a higher diffusion layer pH had greater *in vitro* dissolution rates and, importantly, faster *in vivo* bioavailability.

The pH of the boundary layer at the surface is termed the pH *microenvironment* (pH_m) and is equal to the pH of a saturated solution of the dissolving solid in water (Serajuddin and Jarowski, 1985). The NWNB equation still governs the dissolution rate, but the solubility value is not that of the solute in the dissolution medium but that in a medium of pH_m. As the distance from the surface of the dissolving solid increases, the pH approaches that of the bulk medium (Figure 6.11).

Figure 6.12 shows the pH across the boundary layer for dissolution of haloperidol as a function of dissolution medium pH (Serajuddin, 2007). The value of pH_m is seen to be around 2–3, irrespective of the pH of the bulk medium, and as a consequence the dissolution rate of the salt was observed to be independent of the pH of the dissolution medium used. A similar effect was noted in Chapter 5 for the dissolution of salicylic acid.

Zannou *et al.* (2007) showed that a tablet formulation of a maleate salt of a drug substance lost potency upon storage under accelerated stress conditions while a capsule formulation of the same salt did not (neither did the

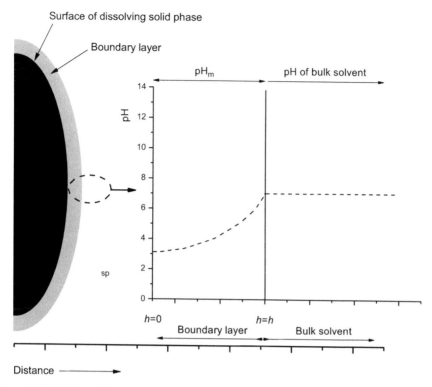

Figure 6.11 Schematic representation of the boundary layer surrounding the surface of a dissolving solid and (inset graph) the change in pH with distance from the surface of the dissolving solid.

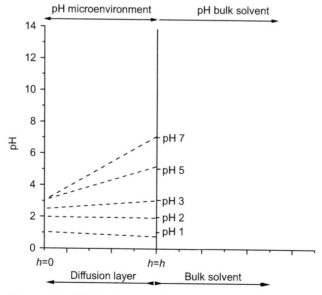

Figure 6.12 pH across the diffusion layer as a function of dissolution medium for haloperidol (redrawn from Serajuddin. Copyright (2007), with permission from Elsevier).

tablet when stored in a capsule shell). The tablet adsorbed water during the storage study, forming a saturated solution on its surface. pH_m was determined to be 4.3, while pH_{max} was 3.3–3.6. Hence the salt converted to the free base upon storage and the free base volatilised under the storage conditions, leading to the loss of potency. The important point here is that a salt can absorb sufficient water to form a saturated solution on its surface. The saturated solution is no different to the boundary layer that arises during dissolution and the effect it can exert on the chemical stability of the salt can be dramatic.

6.4.1 Modification of pH_m

Most solid drug products do not comprise a drug substance alone but contain additional excipients. The question then arises as to the effect of the excipients on the dissolution rate. Obviously, when an excipient acts as a disintegrant, the effect would be to increase the dissolution rate, since the surface area of the dissolving solid would increase. What would happen if dissolution of the excipient caused a change in pH of the solution? Consideration of the NWNB model as well as the concept of a pH microenvironment in the immediate vicinity of the dissolving solid leads to the idea that dissolution rates might be affected.

Table 6.10 shows the pH_m of a number of typical tablet excipients. The pH_m values range from 2.2 to 7.5, which implies that improper selection of

Table 6.10 Measured pH microenvironment of a number of excipients.

Excipient	pH
Dibasic calcium phosphate anhydrous USP[a]	2.21–3.59
PVP[b]	3.7
Microcrystalline cellulose[a,b]	4.03–4.7
Lactose monohydrate[a]	4.24
Mannitol[b]	4.7
Sodium starch glycolate[a]	4.77
Crospovidone[b]	4.9
Colloidal SiO_2[b]	5.2
Sodium croscarmellose[c]	5.7
Lactose[b]	6.1
Maize starch[b]	6.3
Calcium carbonate[a]	6.58–8.07
Magnesium stearate[a,b]	7.1–7.45
Hydrogenated castor oil[b]	7.5

[a] Data from Govindarajan *et al.* (2006).
[b] Data from Zannou *et al.* (2007).
[c] Data from Stephenson, Aburub and Woods (2011).

excipients in a tablet formulation might have a significant impact upon dissolution performance should the pH microenvironment be significantly different from pH_{max}.

Delveridine mesylate provides an example of where pH_m affected stability during storage (Rohrs *et al.*, 1999). For tablets formulated with sodium croscarmellose there was a 20% reduction in the extent of dissolution after 60 min following storage of tablets at 40 °C and 75% RH. The loss in dissolution performance was not seen if tablets were protected from moisture upon storage. It was determined that the mesylate salt was converting to the free base on the surface of the tablet. Sodium croscarmellose has a pH microenvironment of 5.7 (Table 6.10). The pH_{max} of the salt was calculated to be around 4, leading to conversion of the salt upon exposure to moisture and formation of a saturated solution at the surface of the tablet. The solubility of the salt was 2238 times that of the free base, leading to the dramatic drop in dissolution.

It follows that inclusion of an excipient to modify the pH microenvironment in favour of dissolution would be a useful formulation approach. Returning to the maleate salt discussed above, Zannou *et al.* (2007) found that by including citric acid in the formulation the pH microenvironment could be reduced to <3 and a shelf life of greater than 3 years was achieved. Similarly, inclusion of citric acid and disodium hydrogen orthophosphate into a solid dispersion of PVP and frusemide increased dissolution in acidic media (Doherty and York, 1989), while citric acid has also been shown to alter the dissolution of pelanserin hydrochloride (Espinoza, Hong and Villafuerte, 2000) and vinpocetine (Nie *et al.*, 2004).

As noted earlier, the solubility of acidic and basic drug substances will vary as the formulation progresses along the GI tract. While this can be an advantage for targeted release, it is potentially disastrous for controlled release formulations. Addition of a pH modifier (usually an acid) to a controlled release matrix tablet can allow constant dissolution rates along the GI tract. Fumaric acid has been shown to achieve this effect for controlled release tablets containing dipyridamole (Siepe *et al.*, 2006), verapamil hydrochloride (Streubel *et al.*, 2000) and oxybutynin hydrochloride (Varma, Kaushal and Garg, 2005). Alkaline excipients (such as MgO, NaOH, KOH and Na_2CO_3) in a controlled release formulation of telmisartan significantly increased dissolution rates in gastrointestinal fluids (Tran, Thanh and Lee, 2008).

6.5 Partitioning of salts

As noted in Chapter 3, ionised species have a low affinity for organic solvents or nonpolar environments. Thus, while solubility may be enhanced by formation of a salt, there is a considerable risk that partitioning into organic

Table 6.11 Log P and solubility data for ibuprofen sodium (data from Sarveiya et al. (2004)).

pH	Solubility (mg mL^{-1})	log P	% Ionisation
4	0.028	n/d	73.81
5	0.156	3.28	21.98
6	1.0	2.42	2.74
7	340.51	0.92	0.28
8	299.04	0.63	0.03

phases will decrease and hence the *in vivo* permeability of the drug substance is likely to reduce. As ever, there is a compromise to be reached between increasing solubility while maintaining bioavailability and it may well be the case that the most soluble salt is not taken forward for development.

This also highlights another potential drawback of *in vitro* dissolution tests: that optimisation of a drug substance by focusing on improvement in solubility, while improving dissolution rates, does not guarantee improved bioavailability, because permeability is not a factor in conventional dissolution tests. Creating dissolution tests that are a more realistic indicator of *in vivo* performance is critical. Methods to improve IVIVC of dissolution tests have been discussed by Li *et al.* (2005a) and include a combination of physicochemical (solubility, pK_a, salt formation, pH$_m$, particle size and physical form), biopharmaceutical (effect of ionisation and log P on permeability) and physiological (GI tract content, pH and transit time) factors. If a species can ionise then the degree of ionisation will vary with pH (in accordance with the Henderson–Hasselbalch equation). An increased degree of ionisation will decrease partitioning; example data for partitioning of the sodium salt of ibuprofen are given in Table 6.11. Similarly, an increase in log P should increase partitioning. Different salts will have different log P values; data for different salt forms of ibuprofen are given in Table 6.12 that illustrate this effect. Similarly, diclofenac sodium has been shown to have the greatest

Table 6.12 Log P and flux values (across a polydimethylsiloxane model membrane) for various salts of ibuprofen (data from Sarveiya et al. (2004)).

Salt	log P	Flux (μg cm^{-2} h^{-1})
Sodium	0.92	3.09
Ethylamine	0.967	5.42
Ethylenediamine	1.11	15.31
Diethylamine	1.12	7.91
Triethylamine	1.18	48.4

permeability through a model silicone membrane of a range of diclofenac salts (Maitani, Kugo and Nagai, 1994).

Since increasing the hydrophilicity of a drug substance should affect permeability, the preparation of salts has been used as a method for targeting drugs to the colon, since absorption across the intestinal wall can potentially be reduced. For instance, the sulphate ester sodium salt of prednisolone has been suggested as a colon-specific pro-drug of prednisolone (the salt being hydrolysed by sulphatase in colonic media to produce prednisolone; Doh et al., 2003). Similar strategies have been used for colon-specific targeting of many drug substances, including budesonide (budesonide-β-D-glucuronide; Cui, Friend and Fedorak, 1994), dexamethasone (dexamethasone-β-D-glucuronide; Haeberlin et al., 1993) and sulphasalazine (Klotz, 1985).

Summary box 6.4

- Dissolution rate is governed by solubility of the solute in the dissolution medium.

- Salts increase the dissolution rate by virtue of having increased solubility.

- Solubility is pH dependent. The pH that defines the dissolution rate is that of the boundary layer surrounding the dissolving solid (pH_m), not that of the bulk medium.

- The value of the pH_m can be manipulated by selection of excipients. If values of pH_m and pH_{max} are significantly different salt may convert to free base, leading to change in the dissolution rate or loss in potency.

6.6 Summary

Preparation of a salt is a strategy often used to improve aqueous solubility of a drug substance. Other reasons for preparation of a salt include ease of isolation, increasing stability or as a method of purification. A salt co-former must be selected and the salt form to be developed identified early in preformulation. As a general rule, a difference in pK_a of 3 between the free drug substance and the co-former will result in salt formation. A drug substance with a pK_a below 5 will generally not form a salt that will be stable at physiological pH. Salt screens will identify suitable co-formers. The increase in solubility of a salt, over that of the free drug substance, is a result only of the change in pH upon dissolution. As such, salts show no increase in solubility in buffered media (although the rate of dissolution may increase). Because salts can change pH upon dissolution, the pH of the boundary layer surrounding the surface of a dissolving solid may differ significantly from that of the bulk solvent, but ultimately it is the pH of the boundary layer that controls the dissolution rate.

References

Bandura, A.V. and Lvov, S.N. (2005) The ionization constant of water over wide ranges of temperature and density. *J. Phys. Chem. Ref. Data*, **35**, 15–30.

Berge, S.M., Bighley, L.D. and Monkhouse, D.C. (1977) Pharmaceutical salts. *J. Pharm. Sci.*, **66**, 1–19.

Bhogala, B.R., Basavoju, S. and Nangia, A. (2005) Tape and layer structures in cocrystals of some di- and tricarboxylic acids with 4,4′-bipyridines and isonicotinamide. From binary to ternary cocrystals. *Crys. Engng Commun.*, **7**, 551–562.

Bogardus, J.B. and Blackwood, R.K. (1979) Solubility of doxycycline in aqueous solution. *J. Pharm. Sci.*, **68**, 188–194.

Brittain, H.G. (2007) Strategy for the prediction and selection of drug substance salt forms. *Pharm. Technol.*, **31**, 78–88.

Brittain, H.G. (2008) Introduction and overview to the preformulation development of solid dosage forms. In *Preformulation in Solid Dosage Form Development* (eds M.C. Adeyeye and H.G. Brittain), Informa Healthcare, New York. ISBN 0-8247-5809-9.

Childs, S.L., Stahly, G.P. and Park, A. (2007) The salt-cocrystal continuum: the influence of crystal structure on ionization state. *Molec. Pharm.*, **4**, 323–338.

Cui, N., Friend, D.R. and Fedorak, R.N. (1994) A budesonide prodrug accelerates treatment of colitis in rats. *Gut*, **35**, 1439–1446.

Doh, M.J., Jung, Y.J., Kim, I.H. *et al.* (2003) Synthesis and *in vitro* properties of prednisolone 21-sulfate sodium as a colon-specific prodrug of prednisolone. *Arch. Pharm. Res.*, **26**, 258–263.

Doherty, C. and York, P. (1989) Microenvironmental pH control of drug dissolution. *Int. J. Pharm.*, **50**, 223–232.

Espinoza, R., Hong, E. and Villafuerte, L. (2000) Influence of admixed citric acid on the release profile of pelanserin hydrochloride from HPMC matrix tablets. *Int. J. Pharm.*, **201**, 165–173.

Fini, A., Fazio, G., Fernández Hervás, M.-J. *et al.* (1996) Factors affecting the dissolution of diclofenac salts. *Eur. J. Pharm. Sci.*, **4**, 231–238.

Govindarajan, R., Zinchuk, A., Hancock, B. *et al.* (2006) Ionization states in the microenvironment of solid dosage forms: effect of formulation variables and processing. *Pharm. Res.*, **23**, 2454–2468.

Guerrieri, P., Jarring, K. and Taylor, L.S. (2010) Impact of counterion on the chemical stability of crystalline salts of procaine. *J. Pharm. Sci.*, **99**, 3719–3730.

Haeberlin, B., Rubas, W., Nolen, H.W. and Friend, D.R. (1993) *In-vitro* evaluation of dexamethasone-beta-D-glucuronide for colon-specific drug-delivery. *Pharm. Res.*, **10**, 1553–1562.

Heading, R.C., Nimmo, J., Prescott, L.F. and Tothill, P. (1973) The dependence of paracetamol absorption on the rate of gastric emptying. *Br. J. Pharmac.*, **47**, 415–421.

Huang, L.-F. and Tong, W.-Q. (2004) Impact of solid state properties on developability assessment of drug candidates. *Adv. Drug. Del. Rev.*, **56**, 321–334.

ICH Guideline Q3C(R5) (2011) *Impurities: Guideline for Residual Solvents.*

Klotz, U. (1985) Clinical pharmacokinetics of sulphasalazine, its metabolites and other prodrugs for 5-aminosalicylic acid. *Clin. Pharmacokinet.*, **10**, 285–302.

Kramer, S.F. and Flynn, G.L. (1972) Solubility of organic hydrochlorides. *J. Pharm. Sci.*, **61**, 1896–1904.

Kumar, L., Amin, A. and Bansal, A.K. (2008) Salt selection in drug development. *Pharm. Technol.*, **32** (3), 128–142.

Lemke, T.L., Williams, D.A., Roche, V.F. and Zito, S.W. (eds) (2008) Foye's Principles of Medicinal Chemistry, 6th edn, Lippincott, Williams and Wilkins, Baltimore, MD. ISBN 978-0-7817-6879-5.

Li, S.F., He, H.D., Parthiban, L.J. *et al.* (2005a) IV–IVC considerations in the development of immediate-release oral dosage form. *J. Pharm. Sci.*, **94**, 1396–1417.

Li, S., Wong, S.M., Sethia, S. *et al.* (2005b) Investigation of solubility and dissolution of a free base and two different salt forms as a function of pH. *Pharm. Res.*, **22**, 628–635.

Maitani, Y., Kugo, M. and Nagai, T. (1994) Permeation of diclofenac salts through silicone membrane – a mechanistic study of percutaneous-absorption of ionizable drugs. *Chem. Pharm. Bull.*, **42**, 1297–1301.

Morris, K.R., Fakes, M.G., Thakur, A.B. *et al.* (1994) An integrated approach to the selection of optimal salt form for a new drug candidate. *Int. J. Pharm.*, **105**, 209–217.

Nelson, E. (1957) Solution rate of theophylline salts and effects from oral administration. *J. Am. Pharm. Soc. Sci. Ed.*, **46**, 607–614.

Nie, S., Pan, W., Li, X. and Wu, X. (2004) The effect of citric acid added to hydroxypropyl methylcellulose (HPMC) matrix tablets on the release of vinpocetine. *Drug Dev. Ind. Pharm.*, **30**, 627–635.

Patel, A., Jones, S.A., Ferro, A. and Patel, N. (2009) Pharmaceutical salts: a formulation trick or a clinical conundrum? *Br. J. Cardiol.*, **16**, 281–286.

Pudipeddi, M., Serajuddin, A.T.M., Grant, D.J.W. and Stahl, P.H. (2011) Solubility and dissolution of weak acids, bases and salts. In *Handbook of Pharmaceutical Salts. Properties, Selection and Use* (eds P.H. Stahl and C.G. Wermuth), revised 2nd edn, Wiley-VCH. ISBN 3-906-39051-9.

Remenar, J.F., MacPhee, J.M., Larson, B.K. *et al.* (2003) Salt selection and simultaneous polymorphism assessment via high-throughput crystallization. The case of sertraline. *Org. Proc. Res. Dev.*, **7**, 990–996.

Rohrs, B.R., Thamann, T.J., Gao, P. *et al.* (1999) Tablet dissolution affected by a moisture mediated solid-state interaction between drug and disintegrant. *Pharm. Res.*, **16**, 1850–1856.

Russell, T.L., Berardi, R.R., Barnett, J.L. *et al.* (1994) pH related changes in the absorption of dipyridamole in the elderly. *Pharm. Res.*, **11**, 136–143.

Sarveiya, V., Templeton, J.F. and Benson, H.A.E. (2004) Ion-pairs of ibuprofen: increased membrane diffusion. *J. Pharm. Pharmacol.*, **56**, 717–724.

Serajuddin, A.T.M. (2007) Salt formation to improve drug solubility. *Adv. Drug Del. Rev.*, **59**, 603–616.

Serajuddin, A.T.M. and Jarowski, C.I. (1985) Effect of diffusion layer pH and solubility on the dissolution rate of pharmaceutical acids and their sodium salts. II: salicylic acid, theophylline and benzoic acid. *J. Pharm. Sci.*, **74**, 148–154.

Siepe, S., Lueckel, B., Kramer, A. *et al.* (2006) Strategies for the design of hydrophilic matrix tablets with controlled microenvironmental pH. *Int. J. Pharm.*, **316**, 14–20.

Stahl, P.H. and Wermuth, C.G. (eds) (2011) *Handbook of Pharmaceutical Salts. Properties, Selection and Use*, revised 2nd edn, Wiley-VCH. ISBN 3-906-39051-9.

Stephenson, G.A., Aburub, A. and Woods, T.A. (2011) Physical stability of weak bases in the solid-state. *J. Pharm. Sci.*, **100**, 1607–1617.

Streubel, A., Siepmann, J., Dashevsky, A. and Bodmeier, R. (2000) pH-independent release of a weakly basic drug from water-insoluble and -soluble matrix tablets. *J. Cont. Rel.*, **67**, 101–110.

Tarsa, P.B., Towler, C.S., Woollam, G. and Berghausen, J. (2010) The influence of aqueous content in small scale salt screening – improving hit rate for weakly basic, low solubility drugs. *Eur. J. Pharm. Sci.*, **41**, 23–30.

Teraoka, R., Otsuka, M. and Matsuda, Y. (1993) Effects of temperature and relative humidity on the solid-state chemical stability of ranitidine hydrochloride. *J. Pharm. Sci.*, **82** (1993) 601–604.

Tran, P.H.L., Thanh, H.T. and Lee, B.-J. (2008) Modulation of microenvironmental pH and crystallinity of ionizable telmisartan using alkalizers in solid dispersions for controlled release. *J. Cont. Rel.*, **129**, 59–65.

Varma, M.V.S., Kaushal, A.M. and Garg, S. (2005) Influence of micro-environmental pH on the gel layer and release of a basic drug from various hydrophilic matrices. *J. Cont. Rel.*, **103**, 499–510.

Wells, J.I. (1988) *Pharmaceutical Preformulation. The Physicochemical Properties of Drug Substances*. John Wiley & Sons, Ltd, Chichester. ISBN 0-470-21114-8

Yamada, I., Goda, T., Kawata, M. and Ogawa, K. (1990) Application of gastric acidity-controlled beagle dog to bioavailability study of cinnarizine. *Yakugaku Zasshi*, **110**, 280–285.

Zannou, E.A., Yatindra, Q.J., Joshi, M. and Serajuddin, A.T.M. (2007) Stabilization of the maleate salt of a basic drug by adjustment of microenvironment pH in solid dosage form. *Int. J. Pharm.*, **337**, 210–218.

Answers to study questions

6.1 It follows that if salts are dissolved in buffered media there will be no difference in the solubility profile of the salt relative to the corresponding free acid or base, because the buffer will act to neutralise any change in pH. It should also be remembered that the common ion effect may actually reduce the solubility of salt forms relative to the free drug.

6.2 In part this is because the pK_a of hydrochloric acid is so low that there is a good chance that it will form a stable salt with a weak base. Hydrochloride salts are also widely understood and form physiologically common ions and so are acceptable from a regulatory perspective. However, they do have some disadvantages, including the fact that the drop in pH upon dissolution may be significant, causing undesirable local effects, such as irritation upon injection for parenteral formulations. There are also risks of corrosion of manufacturing plant and equipment (i.e. stainless steel components), instability during storage (especially if the salt is hygroscopic) and reduced dissolution and solubility in physiological fluids because of the common ion effect.

7
Physical Form I – Crystalline Materials

7.1   Introduction

Selection of the solid state is probably the most important factor when considering development of a solid drug product, once the drug substance has been identified. Many solid state (or physical) forms may exist (it is important to note that more forms may exist than have actually been discovered) and each will have different physicochemical properties (including solubility, dissolution rate, surface energy, crystal habit, stability, hygroscopicity, strength, flowability and compressibility). In addition, physical forms are patentable, so knowing all of the available forms of a drug substance is essential both in terms of optimising drug product performance but also in ensuring market exclusivity.

7.2   Crystal formation

Crystals form when molecules condense into an ordered pattern. For ordering to occur, molecules have to move (i.e. they must possess a high degree of mobility); these conditions occur when the sample is molten, in solution (although the concentration must be supersaturated), in an amorphous matrix (see Chapter 8) or in the vapor phase. In addition, a metastable crystal may recrystallise to a more stable form with time, an increase in temperature or upon addition of a plasticiser.

Crystal formation can be considered as the sum of two processes, *nucleation* and *growth*. Nucleation means the formation of a small mass (or nucleus) on to which further molecules can condense. Nuclei can form either

Essentials of Pharmaceutical Preformulation, First Edition. Simon Gaisford and Mark Saunders.
© 2013 John Wiley & Sons, Ltd. Published 2013 by John Wiley & Sons, Ltd.

through a small number of molecules becoming aligned (primary nucleation) or by addition of a seed crystal (secondary nucleation). The time taken for nuclei to form via primary nucleation can be considerable and is often the rate-limiting step in crystallisation. Addition of seed crystals obviates this delay and usually leads to rapid crystallisation. In cases where the sample can exist in more than one crystalline form, the seed crystal can be used to select which form crystallises. Sharp edges or rough surfaces in packaging materials or containers may also act as nucleation points, reducing the stability of certain products. Precipitation from solutions for injection is a particular hazard that must be avoided. The shelf-life for etoposide, supplied as a nonaqueous concentrate in a sealed vial, is more than 2 years. However, when diluted for intravenous injection with saline, the shelf life reduces to 24 h at 15–25 °C; storage for longer periods post dilution or in a fridge is not permitted, because of the significant risk of precipitation. Growth is the condensation (via deposition) of further molecules on to the nucleus.

7.2.1 Crystal formation from the melt

Consider a pure compound existing at a temperature just greater than its melting temperature;[1] it will be in a molten, liquid state (assuming sublimation does not occur). The molecules will have a high degree of mobility (moving via rotation, vibration or translation and being essentially randomly oriented) and the system will have a certain enthalpy. Now consider what will happen to the molecules if the temperature of the sample is reduced. A reduction in temperature means a reduction in enthalpy and so the mobility of the molecules will decrease. As the molecules are moving around less there will be a concomitant reduction in volume. Both variables should change linearly with temperature (at least over small temperature ranges). When the temperature is reduced to the melting temperature there will be a *phase transformation* as the material condenses to form a solid phase. The phase change will be characterised by a step (or discontinuity) in enthalpy, because the solid state has a distinct, and lower, enthalpy than the liquid state. Similarly, there will be a reduction in volume because formation of a solid phase requires the molecules to pack closely together and remain closely packed. These processes are represented in Figure 7.1 (Glicksman, 2011). Note from this plot therefore that one characteristic of the crystal form is that it has a defined melting temperature.

---

[1]Melting typically occurs over a range of temperatures, typically 1–2 °C, but for the purpose of this discussion, the ideal of a single melting temperature will be assumed.

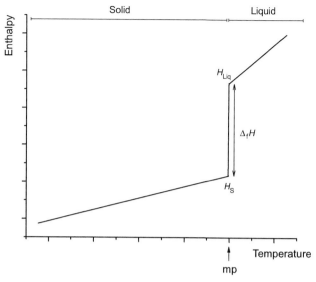

Figure 7.1 Thermodynamic representation of the change in enthalpy as a material crystallizes from a liquid to form a solid phase (reproduced from Glicksman (2011), with kind permission from Springer Science + Business Media B.V.).

7.2.2 Crystal growth from solution

Crystals may grow from solution but this requires that the initial concentration be supersaturated (i.e. above its solubility).

Study question 7.1   Why must the solution be supersaturated for crystallisation to occur?

This may take some time and so the process is often accelerated by adding a seed crystal or other nucleation point (as noted above), reducing the volume of solvent (via evaporation), changing the pH, adding a miscible anti-solvent (i.e. a solvent in which the solute has poor solubility) or reducing the temperature (which usually reduces the solubility of the solute).

The process of crystal growth from a supersaturated solution can be considered to be the reverse of the process of dissolution discussed earlier, involving diffusion of solute molecules to the crystal surface and condensation, and so can be described in similar terms to the Noyes–Whitney equation introduced in Chapter 5. The driving force is still a concentration difference, now between saturation (S_t) and supersaturation (S_s):

$$\frac{dm}{dt} = \frac{DA}{h}(S_s - S_t) \tag{7.1}$$

The remaining terms are the same as defined earlier. In practice crystals often dissolve faster than they grow and so Equation (7.1) is not always found to be valid. An alternative model considers a growth constant (k_g) and an 'order' of crystal growth, n (Mullin, 2001):

$$\frac{dm}{dt} = Ak_g \left(S_s - S_t\right)^n \tag{7.2}$$

In cases where multiple crystal forms may exist, selection of crystallisation solvents may be used preferentially to select one form over another. Other solutes may be added to the solution to change the shape of the crystal produced (see below).

Summary box 7.1

- Crystals form when molecules condense from the liquid state or from a supersaturated solution.

- Crystal growth proceeds via deposition of molecules on to a nucleus.

- The nucleus can be formed in situ by alignment of molecules (primary nucleation) or seed crystals can be added (secondary nucleation).

7.3   Crystal structure

When molecules condense into a solid phase they will usually (but not always – see Chapter 8) orient themselves so that they are in structural alignment, forming a *crystal lattice*. Assuming that the arrangement is such that the enthalpy and volume are minimised, a perfect crystal will have formed. *It is not possible to arrange the constituent molecules of a perfect crystal in any other way that would result in a lower enthalpy or volume.*

The structure of a crystal lattice may be defined by considering the smallest structural arrangement of its constituent molecules that repeats in three dimensions: the *unit cell*. Surprisingly, given the myriad molecules in nature, there are only a limited number of unit cell arrangements. Bravais (1850) first worked out the lattice arrangements and showed there are only seven basic unit cells (Table 7.1). The unit cells are shown diagrammatically in Figure 7.2; the molecules in the unit cells are represented as point sources (the black spheres) and in the basic unit cells appear only at the external corners (called *primitive centring*).

In addition to primitive centring, molecules may also be located at the centre of the unit cell (*body centred*), at the centre of each face of the unit cell

Table 7.1 Axis and angle rules for the seven Bravais unit cells.

Name	Axis rules	Angle rules
Triclinic	$a \neq b \neq c$	$\alpha \neq \beta \neq \gamma$
Monoclinic	$a \neq b \neq c$	$\alpha = \gamma = 90° \neq \beta$
Orthorhombic	$a \neq b \neq c$	$\alpha = \beta = \gamma = 90°$
Tetragonal	$a = b \neq c$	$\alpha = \beta = \gamma = 90°$
Hexagonal	$a = b \neq c$	$\alpha = \beta = 90°\ \gamma = 120°$
Rhombohedral	$a = b = c$	$\alpha = \beta = \gamma \neq 90°$ and $< 120°$
Cubic	$a = b = c$	$\alpha = \beta = \gamma = 90°$

(*face centred*) or at the centre of a pair of opposing faces of the unit cell (*base centred* – three base centred arrangements are possible). This means there are 42 (7 × 6) possible Bravais lattices, although only 14 are unique (the remainder can be shown to be mathematically equivalent). The 14 unique Bravais lattices are shown in Figure 7.2. Pharmaceutical compounds are typically triclinic, monoclinic or orthorhombic.

The unit cell arrangement then repeats in three dimensions to produce the macroscopic crystal. The gross exterior appearance of a crystal is called its *habit*. The USP lists six basic habits (acicular (needle-shaped), blade, columnar, equant (or isometric, including cubic and spherical), plate and tabular; Figure 7.3), although other terms are frequently encountered (such as hexagonal, octahedral, prismatic and pyramidal). The names are simply descriptive terms for the macroscopic shape of the crystal and are not necessarily related to the geometry of the unit cell. Crystal habit can significantly affect particle and powder bulk properties, discussed in Chapters 10 and 11.

Summary box 7.2

- Crystal structures are described in terms of the minimum arrangement of molecules that repeats in a three-dimensional pattern – the unit cell.

- Pharmaceutical crystals are usually monoclinic, triclinic or orthorhombic.

- The macroscopic shape of the crystal is called the habit.

7.4 Polymorphism

When a compound can crystallise to more than one unit cell (i.e. the molecules in the unit cells are arranged in different patterns) it is said to be *polymorphic* (from the Greek, 'many forms'). The form with the highest

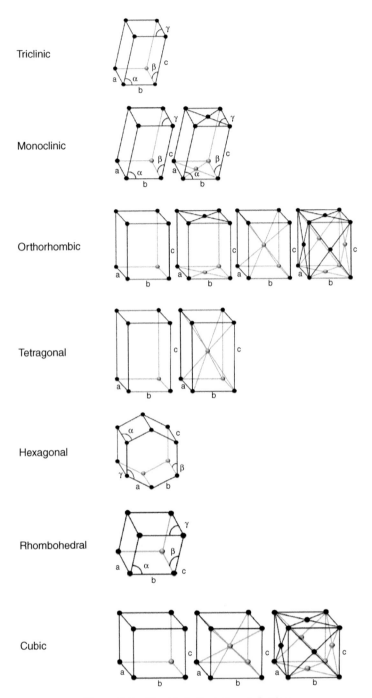

Triclinic

Monoclinic

Orthorhombic

Tetragonal

Hexagonal

Rhombohedral

Cubic

Figure 7.2 The 14 distinct Bravais lattices.

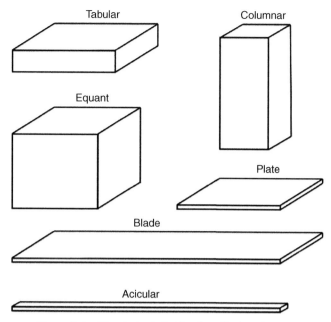

Figure 7.3   The six basic habits described in the USP.

melting temperature is called the *stable* polymorphic form and all other forms
are *metastable*.

The formation of polymorphs can be considered using the same thermodynamic argument adopted earlier. When the rate of cooling is slow
the molecules have sufficient time to organise themselves in the most efficient (enthalpy-minimised) arrangement possible and the stable polymorph
is formed. If the rate of cooling is sufficiently fast that the stable polymorph
does not have time to form, the material will remain in the liquid phase below
the melting temperature of the stable form; this is termed a supercooled liquid. At some lower temperature the molecules will align themselves in an
alternate, but less efficiently packed, crystal form and there will be a discontinuity in enthalpy or volume as a metastable crystal is formed (Figure 7.4).
Thus, the metastable form is characterised as having greater enthalpy and
volume and a lower melting temperature than the stable form.

Study question 7.2   If other metastable polymorphs exist how would Figure 7.4 be extended?

7.4.1   Thermodynamics of polymorphism

So far it has been assumed that the same form is the stable polymorph under
all conditions of temperature (and pressure, but temperature is the more

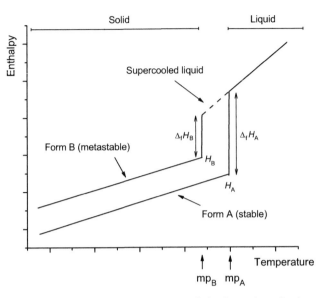

Figure 7.4 Thermodynamic representation of the formation of polymorphs.

important parameter in the context of preformulation). If so, the compound is said to exhibit *monotropic* polymorphism. It is possible, however, that with a change in temperature a different form becomes the stable form. In this case the compound exhibits *enantiotropic* polymorphism.

The concepts of monotropic and enantiotropic polymorphism can be explained with reference to enthalpy and free energy curves plotted against temperature (again, pressure is assumed to be constant, but analogous situations occur with changes in pressure – see Lohani and Grant, 2006). Figure 7.5 shows the case for a monotropic pair of polymorphs. At absolute zero entropy is zero and so enthalpy and free energy are equal. As temperature increases, enthalpy increases and free energy decreases (because entropy will increase) in accordance with the familiar thermodynamic relationship

$$\Delta G = \Delta H - T\Delta S \tag{7.3}$$

Note that although (as assumed in constructing Figure 7.1) the change in enthalpy or free energy might be assumed to be linear over small temperature changes, over large temperature ranges nonlinear behaviour is seen. This is one of the reasons for exercising caution when using Arrhenius extrapolations during stress testing.

The form with the lowest free energy at any particular temperature is the most stable. In the case of the monotropically related pair represented in Figure 7.5 the most stable solid form is A, at all temperatures up to the melting temperature (the point at which the liquid curve falls below the solid

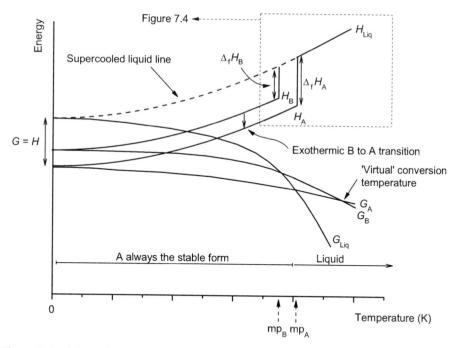

Figure 7.5 Schematic representation of the change in free energy with temperature for two monotropic polymorphs.

curve). Form B exists as a solid form because it has a lower free energy than the liquid at all temperatures below its melting temperature, but is always metastable with respect to form A. The free energy curves for forms A and B eventually reach a point of intersection (at which point form B would become the most stable form), but not until after intersection with the liquid curve (by which point both forms have melted). If form B were to convert to form A, which from a thermodynamic perspective is inevitable, but from a kinetic perspective may take some considerable time, there would be an exothermic change in enthalpy, as form B always has greater enthalpy than form A. Many drugs exhibit monotropic polymorphism, including, for example, most of the barbiturates, the sulphonamides, spironolactone, premafloxacin, furosemide and indomethacin.

The case for an enantiotropically related pair is shown in Figure 7.6. The changes in enthalpy and free energy are analogous to those described above for a monotropically related pair, with the exception that the point of intersection of the free energy curves of the two solid forms occurs before the intersection of either with the liquid curve. Since the form with the lowest free energy is the most stable, before the temperature of intersection form A is stable and form B metastable; after the temperature of intersection the

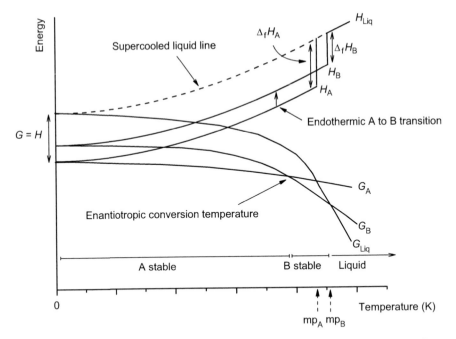

Figure 7.6 Schematic representation of the change in free energy with temperature for two enantiotropic polymorphs.

situation is reversed, form B being stable and form A being metastable. Correspondingly, form A melts at a lower temperature than form B. As the temperature of intersection is exceeded, form A will (from a thermodynamic perspective) convert to form B, now with an endothermic change in enthalpy (as form B always has a higher enthalpy than form A).

This leads to the *heat of transition rule* for determining whether a pair of polymorphs are monotropically or enantiotropically related:

- An exothermic enthalpy of transition is seen for monotropic pairs.
- An endothermic enthalpy of transition is seen for enantiotropic pairs.

Other rules for determining whether a monotropic or an enantiotropic relationship exits are:

Enthalpy of fusion rule. When the higher melting polymorph of a pair also has the higher enthalpy of fusion, they are monotropically related.

Entropy of fusion rule. When the higher melting polymorph of a pair also has the higher entropy of fusion, they are enantiotropically related.

Heat capacity rule. When the higher melting polymorph of a pair also has the higher heat capacity, they are enantiotropically related.

Solubility rule. When the higher melting form of a pair also has the greatest solubility (at temperatures above the transition temperature), the pair are enantiotropically related.

Examples of drugs that exhibit enantiotropic polymorphism include neotame, albendazole and sulphathiazole.

Summary box 7.3

- If the same compound can form crystals with different unit cells it is said to be polymorphic.

- The polymorph with the highest melting temperature is the stable form and all others are metastable.

- With time, accelerated with temperature and/or plasticisers, metastable forms will transform to the stable form.

- If one form is stable at all temperatures (and pressures) the compound exhibits monotropic polymorphism.

- If different forms are stable at different temperatures (and pressures) the compound exhibits enantiotropic polymorphism.

7.4.2 Physicochemical properties of polymorphs

Figure 7.7 (top) shows a schematic representation of two polymorphic forms of a drug. Although nothing of the chemical structure or nature of the drug is specified it is apparent from the diagram that the two polymorphs probably exhibit different physical properties; one that is evident is density, with form I appearing to have a greater number of molecules per unit volume. Since the molecules are more closely packed, form I is also likely to have a greater lattice energy (strength between molecules) and hence a higher melting point than form II.

Study question 7.3 What other properties might change with physicochemical form?

A defining feature of polymorphs is thermodynamic stability (in this case referring to the physical form itself rather than any chemical decomposition or interaction of its constituent molecules). Since only the stable form can be considered to be at a position of thermodynamic equilibrium, over time

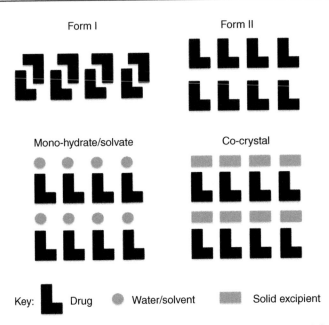

Figure 7.7 Schematic representation of polymorphic forms of a drug (top left and right) as well as a monohydrate or solvate (bottom left) and a co-crystal (bottom right).

the metastable forms will eventually convert to the stable form. It is tempting therefore to consider formulating only the stable polymorph of a drug substance, since this ensures that there can be no change in the polymorph upon storage.

If the stable polymorph shows acceptable bioavailability then it is of course the best option for development (patent issues aside). However, sometimes the stable form has poor processing characteristics (the stable form I of acetaminophen, for instance, has poor compressibility, unlike the metastable form II). Moreover, with the increasing number of poorly soluble drug substances being developed, formulation of a metastable form can be an attractive option for increasing bioavailablity, the classic example being provided by chloramphenicol palmitate (Figure 7.8), where the β form shows much greater bioavailability than the α form (and so there are strict limits in the pharmacopoeia monograph on the amount of the α form permissible in the drug product). Selection of the most appropriate polymorph for development is therefore often a compromise between increased bioavailability and decreased physical stability.

A further consideration is that it is never possible to know *absolutely* that the stable form has been identified.

Study question 7.4　Why is this so?

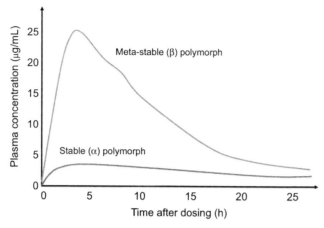

Figure 7.8   Blood plasma concentrations versus time for two polymorphs of chloramphenicol palmitate (redrawn from Aguiar *et al.* (1967), with permission from John Wiley & Sons, Inc.).

How to determine the rate of conversion? The data presented in Figures 7.5 and 7.6 are based on thermodynamics alone and so provide no kinetic information, yet it is this rate that defines whether a particular polymorph is suitable for formulation or not. The only option is to conduct stability trials, in the same way that chemical stability profiles are determined (discussed in Chapter 9), using DSC, XRPD or any other suitable spectrophotometric technique as an assay (looking for a change in physical form, rather than a change in chemical composition). Polymorph conversion will proceed faster with an increase in temperature (and often humidity) and so accelerated conditions may be used to shorten the experimental study time.

7.5   Pseudopolymorphism

The concepts of unit cells and polymorphism have assumed only one molecular species in the system. However, some drugs are well known to form crystal structures containing stoichiometric ratios of other components (co-formers). When the co-former is water the crystal is a *hydrate* (monohydrate when the ratio is 1, dihydrate for 2 and so on). When the co-former is a solvent, the crystal is a *solvate* (methanolate, ethanolate, etc). When the co-former is ordinarily a solid at room temperature and pressure the crystal is a *co-crystal*. These arrangements are shown schematically in Figure 7.7 (bottom).

The presence of pseudopolymorphs and their misidentification as true polymorphs is a particular danger during preformulation when (i) the compound is unfamiliar and not well characterised and (ii) solvents may have been used during synthesis or purification. Pseudopolymorphs will have different physicochemical properties and in the absence of any data that suggest otherwise can easily be believed to be true polymorphs. To assign them as

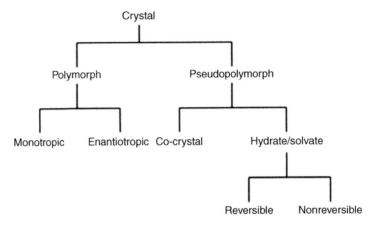

Figure 7.9   Overview of crystalline forms.

such is dangerous, however, not only because they contain at least one additional substance that may have a pharmacological effect but also because they may be unstable with time, temperature and/or relative humidity (RH). Crystal structures often lose water of hydration with increasing temperature or decreasing RH to form a *dehydrate* (effectively a dehydrated unit cell) – if the unit cell is stable in the dehydrate form then the hydrate may well reform if the temperature is reduced or the RH increased (in which case the hydrate is said to be *reversible*). If the dehydrate unit cell is unstable, then rearrangement to an anhydrous form may occur (in other words, a unit cell that does not contain point spaces where the water molecules should be), in which case the hydrate may not reform (and so is termed *irreversible*). The same concept applies to solvates and the terms *desolvate* and *ansolvate* are used.

The range of solid-state forms can thus be summarised as in Figure 7.9.

Summary box 7.4

- If a second component (a co-former) is present in the unit cell the crystal is a pseudopolymorph.

- The co-former can be:

  - water (hydrate),

  - solvent (solvate),

  - any other compound that is a solid at room temperature and pressure (co-crystal).

7.6 Polymorph screening

Polymorph screening is performed in much the same manner as described earlier for salt screening. Basic screening is achieved by crystallising the drug substance from a number of solvent or solvent mixtures of varying polarity. A small amount of drug substance (ca. 0.5 mg) is added into each well of a 96-well plate. To each well is added a small volume of each solvent or solvent mixture.

After an appropriate length of time, the presence in each well of crystals is checked with an optical device (for instance a microscope or a nephelometer). If no crystals are seen then the plate can be stored at a lower temperature. If the reduction in temperature does not cause precipitation then as a last attempt the temperature can be increased to evaporate the solvent.

XRPD can be used to record a diffractogram of the crystals in each well. As noted earlier, care must be taken when interpreting the data that hydrates and solvates are not incorrectly identified as true polymorphs. Visualising the crystals under a hot-stage microscope, fitted with cross-polarised filters, is an alternative option.

DSC also offers the potential for screening polymorphic forms if the compound can form a glass. The underlying experimental principle for formation of metastable forms is based on Ostwald's rule of isolation in stages, which posits that when a material crystallises from a nonequilibrium, high-energy state (such as a glass) it will do so via progression through any available lower energy states; the physical manifestation of this is that the sample will crystallise in a sequence, progressing through any metastable polymorphs to the stable crystalline form. Thus, a sample can be quench-cooled in the DSC to form a glass. Upon heating, the sample may crystallise to a metastable form and melt (Figure 7.10 shows the case for paracetamol). Alternatively, where no crystallisation is seen upon heating, the sample can be annealed to allow crystallisation to the (least stable) polymorphic form.

7.7 Characterisation of physical form

Absolute characterisation of physical form requires determination of the dimensions and positions of the molecules in the unit cell and can be achieved only with single crystal X-ray diffraction. Full structural analysis thus requires a crystal to be grown (at least 0.5–1 mm along its shortest edge) and considerable run-time on a diffractometer. While not unreasonable, preformulation is taken here to mean preliminary characterisation of a compound (probably several structurally related compounds) with a view to selecting the best candidate for development. As such, full structural determination would probably occur later in development and so will not be considered here. X-ray powder diffraction (XRPD), however, is a very useful tool for the

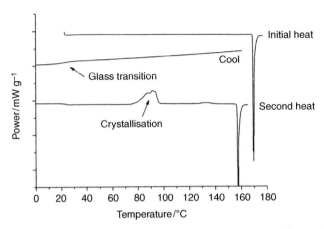

Figure 7.10   DSC thermal traces showing the melting of paracetamol form I (initial heat), quench cooling to form a glass (cool) and crystallisation to and melting of the metastable form II (second heat) (data courtesy of Asma Buanz).

preformulation scientist, requiring little sample (one or two powder parti-cles). No sample pretreatment is required and the measurement is nonde-structive, meaning the sample can be recovered and used in further analyses. In addition, XRPD can make measurements as a function of temperature and RH, which is extremely useful for determining stability of forms. Data are presented as a diffractogram, which plots intensity as a function of beam angle (2θ). A particular form will have a unique number of peaks and so different physical forms are easily identified. Figure 7.11 shows the diffrac-togram for sulphapyridine.

DSC is also a very useful instrument for characterising polymorphic forms. DSC requires no sample preparation, uses small amounts of sample (0.5–1 mg) and provides information on the thermodynamics of phase transitions. It does suffer the drawback that samples may be irreversibly changed (or degraded) after heating, but equally it will detect any phase transitions that the sample might undergo. The combination of XRPD and DSC data is extremely powerful and provides the basic information with which identification and assignment of physical form can be made. Other analytical tools, such as near infrared (NIR), Fourier transform infrared (FT-IR) or Raman spectroscopy, dynamic vapour sorption (DVS), thermo-gravimetry (TGA) and imaging techniques, such as hot-stage and scanning electron microscopy (SEM) provide supplementary data that help confirm the initial interpretation.

7.7.1   Characterisation of polymorphs

If available, XRPD provides the best method to identify and differen-tiate polymorphs. Figure 7.11 shows the powder diffractograms for two

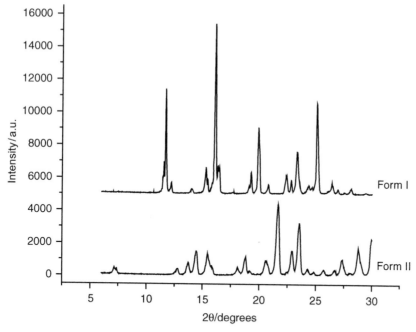

Figure 7.11 XRPD diffractograms for two polymorphs of sulphapyridine (data courtesy of Asma Buanz)

polymorphs of sulphapyridine; it is immediately apparent that each has a unique set of intensity peaks and so the forms are quantitatively different. The 2θ angles for each peak provide a 'fingerprint' for each form. Assuming that the instrument was used in the reflection mode and the sample was spun during the measurement, then the areas of each peak can be used as the basis for a quantitative assay for each form (peak intensities are not used because they can vary with habit or particle size).

Unlike XPRD, DSC data provide no information on structural alignment but differentiate polymorphs on the basis of melting point and enthalpy of fusion, thus providing thermodynamic information instead. This means that DSC can identify which polymorph is stable and which is metastable (which XRPD data cannot) and whether two polymorphs are monotropically or enantiotropically related. In addition, the enthalpy of fusion can be used to calculate ideal solubility.

Assuming that there is only one polymorph in a sample and that it is the stable form, heating the sample in the DSC should result in a thermal trace showing only an endothermic melt (Figure 7.12) (assuming that no degradation occurs at the melt). From this the melting temperature (either as the extrapolated onset or peak maximum, noting that the position of a peak maximum will be dependent both on sample size and heating rate) and

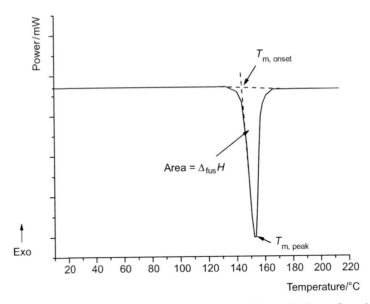

Figure 7.12   Schematic DSC thermal trace showing melt of the stable form of a polymorph.

enthalpy of fusion are easily determined. If the sample put into the DSC initially is a metastable form, then an alternate thermal trace is likely (Figure 7.13, top). Here three events are seen: an endotherm followed by an exotherm followed by an endotherm. To what phase transitions can these events be assigned? The low-temperature endotherm is easily assigned to melting of the metastable form. At a temperature immediately after the endotherm the sample is thus molten, but because the form that melted was metastable, and so at least one higher melting point form is available, the liquid is supercooled with respect to the stable form. With time the liquid will crystallise to the next thermodynamically available solid form (in this case the stable polymorph). Crystallisation is (usually) exothermic and so accounts for the exotherm on the DSC thermal trace. Finally the stable form reaches its melting temperature and melts (the higher temperature endotherm).

This pattern of transitions (endotherm–exotherm–endotherm) is a characteristic indicator of the presence of a metastable polymorph, assuming a monotropic relationship.

Study question 7.5   What would the DSC thermal trace look like if two polymorphs are enantiotropically related (see Figure 7.14)?

Study question 7.6   If more than one metastable form exists and a monotropic relationship exists, how would the DSC thermal trace appear (see Figure 7.14)?

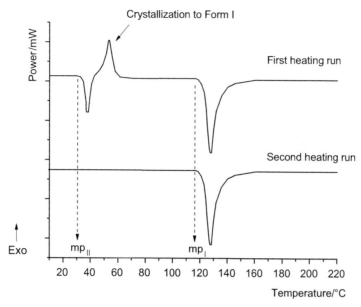

Figure 7.13 Schematic representation of the DSC thermal traces for a metastable polymorph on its first (top) and second (bottom) heating runs.

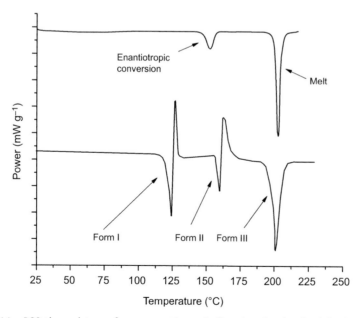

Figure 7.14 DSC thermal trace for an enantiotropically related pair of sulphathiazole polymorphs (top, courtesy of Asma Buanz) and a series of monotropically related premafloxacin polymorphs (bottom, redrawn from Schinzer *et al.* (1997)).

However, as is always the case with DSC, interpretation is at best qualitative and needs additional experimental verification. One option is to cool the sample to room temperature and then reheat. If the cooling is reasonably slow then the stable form will have time to crystallise and as a consequence when reheated the sample will show only one endotherm, corresponding to melt of the stable form (Figure 7.13, bottom).

Another is to repeat the experiment at a (significantly, at least an order of magnitude) faster heating rate. Figure 7.15 shows the thermal traces for a metastable polymorph at slow (top) and fast (bottom) heating rates (the top trace is the same data as in Figure 7.13). It is apparent that the endothermic peaks appear at the same temperatures but the exothermic peak appears at a higher temperature when a faster heating rate is used. The reason for this is because melting and crystallisation are different types of transition.

In progressing through a melt a sample moves from an ordered to a disordered state. The only barrier to melting is the crystal lattice energy that holds the crystal together. Once the sample has sufficient energy to overcome the crystal lattice energy (i.e. at the melting temperature) it will melt. Assuming there are no significant thermal gradients in the sample, then melting can be considered to occur instantaneously once the melting temperature has been

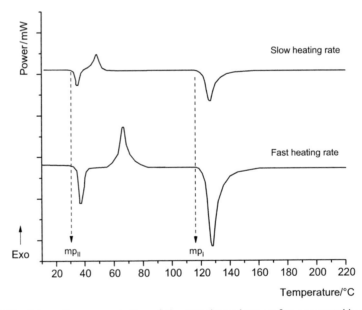

Figure 7.15   Schematic representation of the DSC thermal traces for a metastable polymorph at slow (top) and fast (bottom) heating rates.

reached and, as such, melting endotherms show no dependence upon heating rate (they are termed *thermodynamic* transitions).

Crystallisation, conversely, involves transformation from disorder to order and so as well as requiring a certain amount of energy to transform (i.e. reaching a certain temperature) there is also a finite amount of time required for all the molecules to rearrange into their new form. During this time the instrument is increasing in temperature; the faster the heating rate the greater the rise in temperature before crystallisation occurs. Hence, the temperature at which crystallisation is seen increases with the scan rate (the transition is termed *kinetic* and is scan rate dependent).

Of course, in practice the definitions and theories of thermodynamic and kinetic transitions are more complex than the rather simple overview given above, but the relevant point is that thermodynamic transitions show no dependence upon scan rate while kinetic transitions do and this difference can be used as an aid to data interpretation. In the case of polymorphic transitions, the net result is as shown in Figure 7.15. The two melting endotherms appear at the same temperature while the crystallisation exotherm appears at a higher temperature when a faster heating rate is used.

This has multiple advantages. One, clearly, is in assigning DSC peaks to physical events. Melting events should not move with heating rate while crystallisation events will. Another, not explicitly noted above but implicit from the data in Figure 7.15, is that at a slow heating rate the processes of melting of the metastable form and crystallisation to the stable form overlapped; this makes it difficult to assign an onset temperature to the exotherm and renders it impossible to determine the enthalpy of either process. By heating the sample at a faster rate it is often possible to separate the two transitions (because the crystallisation event will move to a higher temperature) and hence a full thermodynamic analysis of each can be performed. Such an approach has been used to characterise form III carbamazepine (McGregor, Saunders and Buckton, 2004) and two forms of a Merck development compound (McGregor and Bines, 2008).

Extending the principle that thermodynamic events shift to higher temperatures with increasing scan rates, consider what would happen if the heating rate used for the sample in Figure 7.15 were so high that the temperature at which crystallisation would appear became *higher* than the melting temperature of the stable form. *Neither event would occur*. This is because crystallisation to the stable form would not in principle occur until after the melting point of the stable form had been passed (and so crystallisation would not occur in the first place) (Figure 7.16). The heating rate has become fast enough that crystallisation has been *inhibited*. The minimum heating rate required to inhibit a kinetic transition can be determined by plotting the transition temperatures determined at a number of heating rates (Figure 7.17)

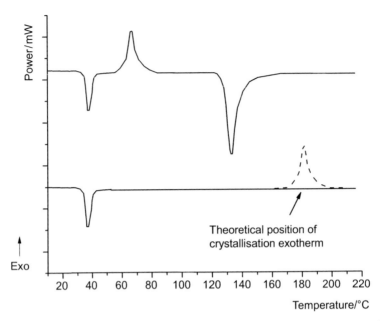

Figure 7.16 Schematic representation of the theoretical position of the crystallisation exotherm when a very fast heating rate is used.

(obviously this approach requires that the higher temperature thermodynamic transition arises as a result of the lower temperature kinetic transition).

Study question 7.7   What other experimental data could you record that would confirm the interpretation of the DSC data above?

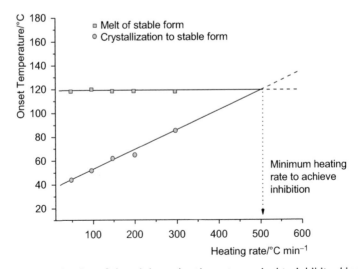

Figure 7.17   Determination of the minimum heating rate required to inhibit a kinetic event.

Summary box 7.5

- XRPD and DSC provide the basic information upon which to identify and characterise polymorphs.

- XRPD provides structural information. DSC provides thermodynamic information.

- Stable polymorphs should show a single melting endotherm by DSC.

- Metastable polymorphs often show a characteristic sequence of phase transitions (endotherm–exotherm–endotherm) corresponding to melt, crystallisation, melt, assuming they are monotropically related.

- Enantiotropically related polymorphs show an endothermic transition.

- Fast heating rates can be used to isolate, and in the limit to inhibit, crystallisation following melt of a metastable form.

7.7.2 Characterisation of pseudopolymorphs

Characterisation of pseudopolymorphs proceeds in the same manner as for polymorphs discussed above, noting that while XRPD data will differentiate between forms they will not necessarily indicate the presence of co-formers in the unit cell. DSC again is capable of discriminative analysis. Taking the case of a hydrate first, a broad endotherm is usually seen (around 100 °C, although this can be higher, sometimes as high as 170 °C, if the water is bound to the drug substance), corresponding to loss of the water of hydration. Following the broad endotherm, there may be an exotherm (corresponding to conversion of the dehydrate to the anhydrate); the presence of the exotherm indicates an irreversible hydrate. Finally there should be an endotherm as the crystal melts. These events are shown in Figure 7.18 for reversible and irreversible hydrates. The case for solvates is analogous, except that the broad endotherm usually occurs at a lower temperature (as the solvent is usually more volatile).

Again, several experimental options are available to aid interpretation. One would be to cool the sample down following initial heating (cooling can commence as soon as the broad endotherm has finished – there is no need to heat through any further events) and then reheat. No broad endotherm should be seen during the second heat (as the water or solvent is no longer present – even if the hydrate or solvate is reversible, it cannot reform since the DSC is purged with a dry gas). Another would be to repeat the experiment

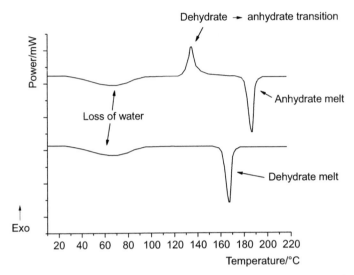

Figure 7.18   Schematic DSC thermal traces for an irreversible hydrate (top) and a reversible hydrate (bottom).

in hermetically sealed and nonhermetically sealed pans. Since evaporation occurs into the headspace of the pan, the shape (and possibly magnitude) of the broad endotherm should be different for the two pans. The final option is to use a very fast heating rate – fast enough that the sample reaches the melting temperature before the water or solvent has a chance to evaporate. Because evaporation is a kinetic event it takes a finite time to occur. If the sample is heated fast enough that it progresses through another transition (melting in this case) before evaporation can occur, the melt will be that of the hydrate or solvate.

Hot-stage microscopy as ever offers useful supporting evidence. As above, cross-polarised filters can be used to confirm form changes with temperature. In addition, hydrates or solvates may be identified if the sample is placed on the slide under a mineral or silicone oil. Upon melting, a hydrate or solvate will be seen to produce bubbles, as the volatile solvent evaporates into the oil.

Earlier it was noted that an endotherm–exotherm–endotherm sequence of transitions is a strong indicator of the presence of a metastable polymorph. Inspection of the data in Figure 7.18 shows a similar sequence of events and yet in that case the sample was a hydrate or solvate.

Study question 7.8   What other measurement could you make to help interpret the DSC data in this case (see Figure 7.19)?

Co-crystals open the intriguing formulation strategy of combining two active compounds in one crystal structure, the physicochemical properties of

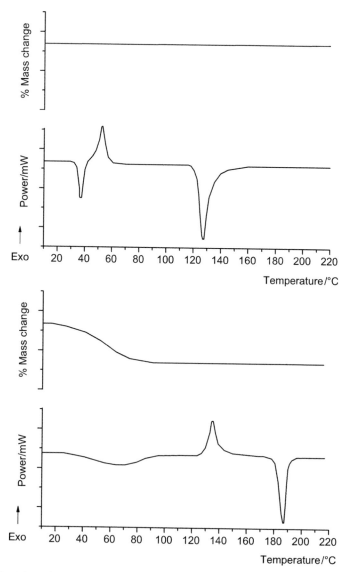

Figure 7.19 DSC and TGA thermal traces for a metastable polymorph (top) and an irreversible hydrate (bottom).

which are more advantageous than those of either of the actives in their pure form. A classic example of this type of formulation is EMLA (a eutectic mixture of local anesthetics) cream, which contains 2.5% each of lidocaine and prilocaine. Typically an improvement in solubility is seen although improvements in hygroscopic properties may also be achieved. In addition, it is possible that formulation as a co-crystal could reduce the impact of polymorphism for highly polymorphic compounds, although this presupposes that fewer polymorphic arrangements are available when two compounds are present in the unit cell.

Co-crystals often do not display the stability issues (with temperature or humidity) seen for hydrates and solvates, principally because the additional component is not volatile (unless it can sublime – theophylline and benzoic acid are good examples). Nor can they be confirmed with hot-stage microscopy (although DSC, spectroscopy or XRPD will detect them). However, co-crystals are extremely unlikely to have been formed accidentally during preformulation in the way solvates and hydrates are, because stoichiometric amounts of the co-former would need to have been present during synthesis or purification.

Summary box 7.6

- Hydrates and solvates typically show loss of water or solvent by DSC. Confirmation is possible with hot-stage microscopy or TGA.

- Crystallisation to an anhydrate or ansolvate form suggests that an irreversible hydrate or solvate was formed initially.

7.8 Summary

Selection of physical form is nearly as important as selection of chemical structure. The latter affects efficacy against a biological target while the former affects, among others, ease of manufacture and isolation, stability, dissolution rate and bioavailability, processability, hygroscopicity and patentability. Unless there is a good reason not to, it is safest to select the stable crystal form for development. If the solubility and dissolution profiles are unsatisfactory, then a metastable or pseudopolymorphic form can be selected, although it is then absolutely necessary to understand the rate of conversion to the stable form with respect to temperature and humidity to ensure there is no loss in drug product performance upon storage. If satisfactory bioavailability still cannot be achieved, then the focus could move to formulation in an amorphous form (Chapter 8).

References

Aguiar, A.J., Krc Jr, J., Kinkel, A.W. and Samyn, J.C. (1967) Effect of polymorphism on the absorption of chloramphenicol from chloramphenicol palmitate. *J. Pharm. Sci.*, **56**, 847–853.

Bravais, A. (1850) Mémoire sur les systèmes formés par les points distribués régulièrement sur un plan ou dans l'espace. *J. Ecole Polytech.*, **19**, 1–128.

Gaisford, S., Buanz, A.B.M. and Jethwa, N. (2010) Characterisation of paracetamol Form III with rapid-heating DSC. *J. Pharm. Biomed. Anal.*, **53**, 366–370.

Glicksman, M.E. (2011) *Principles of Solidification. An Introduction to Modern Casting and Crystal Growth Concepts*, Springer. ISBN 978-1-4419-7343-6.

Grant, D.J.W. (1999) Theory and origin of polymorphism. In *Polymorphism in Pharmaceutical Solids* (ed. H.G. Brittain), Informa Healthcare. ISBN 0-8247-0237-3.

Lohani, S. and Grant, D.J.W. (2006) Thermodynamics of polymorphs. In *Polymorphism in the Pharmaceutical Industry* (ed. R. Hilfiker), Wiley-VCH. ISBN 3-5273-1146-7.

McGregor, C. and Bines, E. (2008) The use of high-speed differential scanning calorimetry (Hyper-DSC) in the study of pharmaceutical polymorphs. *Int. J. Pharm.*, **350**, 48–52.

McGregor, C., Saunders, M. and Buckton, G. (2004) The use of high-speed differential scanning calorimetry (DSC) to study the thermal properties of carbamazepine polymorphs. *Thermochim. Acta*, **417**, 231–237.

Mullin, J.W. (2001) *Crystallization*, 4th edn, Butterworth-Heinemann, Oxford. ISBN 0-7506-4833-3.

Schinzer, W.C., Bergren, M.S. and Aldrich, D.S. (1997) Characterization and interconversion of polymorphs of premafloxacin, a new quinolone antibiotic. *J. Pharm. Sci.*, **86**, 1426–1431.

Answers to study questions

7.1 This is because the position of equilibrium lies between the solid (crystalline) phase and (saturated) solution. Starting from a solution, in order for any solid phase to precipitate, the concentration must be above saturation. If there was excess solid and the solution concentration was less than saturated, then the solid would dissolve until saturation was achieved.

7.2 The supercooled liquid line would be extended to a lower temperature, where again a discontinuity would be seen as another metastable form crystallised. The volume or energy of each metastable form will increase as they become further apart from the stable form.

7.3 Grant (1999) lists a large number of factors that will change with polymorphic form and divides them into groups (Table 7.2). Note also that polymorphic forms are patentable and so ensuring market exclusivity rests on understanding, and patenting, each available form. Atorvastatin, for instance, has 37 known or claimed polymorphs.

7.4 Finding a new form with a higher melting point proves that the lower melting point form is metastable. The definition of the stable form is that no more stable form can exist and so there can be no experimental verification. Although this seems like a philosophical point, it has caused problems for marketed products.

Table 7.2 Properties that may change with polymorphic form (adapted from Grant (1999)).

Group	Property
Packing	Molar volume, density, refractive index, hygroscopicity
Thermodynamic	Melting temperature, heat capacity, enthalpy, entropy, free energy, solubility, thermodynamic activity, vapour pressure
Kinetic	Rates of sublimation, dissolution, solid-state degradation or form conversion
Surface	Interfacial tension, surface free energy, habit
Mechanical	Compressibility, hardness
Spectroscopic	UV, IR, Raman, NMR, XRPD

Ritonavir, an antiretroviral drug developed by Abbott, provides a good example. The initial drug product contained a crystal form with a melting point of 122 °C. After roughly two and a half years on the market, and following the manufacture of 240 batches of the product, a new crystal form of Ritonavir appeared. The new form had a slower dissolution rate, so batches of drug product containing this form failed dissolution release specification and turned out to have a higher melting temperature of 125 °C. In other words, it was a more stable polymorph. The problem facing Abbott was that once the new form appeared it quickly contaminated all of their manufacturing facilities and, despite efforts to remove it, once present they could only produce the more stable form. As a consequence the product (marketed as Norvir) was withdrawn from the market and redeveloped based on the new, more stable, form.

7.5 From Figure 7.6, conversion of polymorphs will occur at the transition temperature and will be endothermic. The DSC thermal trace will appear as shown in Figure 7.14 (top), in this case for an enantiotropically related pair of sulphathiazole polymorphs.

7.6 A series of endothermic (melting) and exothermic (crystallisation) events will be seen, before melting of the stable form. Figure 7.14 (bottom) shows the DSC data for form III premafloxacin converting to form II and then form I. All forms are monotropically related.

7.7 Additional verification of the occurrence of polymorphic changes can be provided with hot-stage microscopy if fitted with cross-polarising filters, because polymorphic forms are birefringent and so appear coloured. A change in polymorphic form with temperature will occur with a change in colour pattern in the microscope. This effect is discussed in more detail in Chapter 8. Thermogravimetric data would show no change in mass during a change in form, which would also provide additional evidence that the DSC data did not arise from loss of a volatile component.

7.8 TGA offers one option. Remembering that TGA is sensitive to changes in sample mass, then hydrates and solvates will show a loss in mass at the temperature of the first endotherm while metastable polymorphs will not (Figure 7.19).

8
Physical Form II – Amorphous Materials

8.1 Introduction

Sugar cubes and candy floss are both forms of sugar. Everyday experience suggests that sugar cubes need to be stirred to aid dissolution, even into a hot liquid such as tea, while candy floss immediately 'melts' in the mouth. The only difference between these forms is that in sugar cubes there is an ordered crystalline arrangement of sugar molecules, while in candy floss the molecules are randomly oriented (or *amorphous*, from the Greek, 'lack of form'). Amorphous materials are high-energy materials without a crystal lattice and hence one of the main barriers to dissolution is removed; as such they usually have appreciably higher solubilities and faster dissolution rates than their crystalline equivalents, and so offer an alternative to salt selection as a strategy to improve the bioavailability of poorly soluble compounds. However, being in a thermodynamically unstable state, amorphous materials will change structure with time, first by relaxation and ultimately by crystallisation. The benefits of formulating in an amorphous form must therefore be set against the risk of change (and hence reduction in performance) upon storage.

8.2 Formation of amorphous materials

In Chapter 7 it was noted that when materials condense to form a solid phase, they usually align into a repeating order. However, several factors can make it difficult for molecules to orient themselves in large numbers into repeating arrays. One is if the molecular weight of the compound is very high

Essentials of Pharmaceutical Preformulation, First Edition. Simon Gaisford and Mark Saunders.
© 2013 John Wiley & Sons, Ltd. Published 2013 by John Wiley & Sons, Ltd.

(if the drug substance is a derivatised polymer or a biological species, for example). Another is where the solid phase is formed very rapidly (say by quench-cooling or precipitation), whereby the molecules do not have sufficient time to align. It is also possible to disrupt a pre-existing crystal structure with application of a localised force (such as during particle size reduction). In any of these cases, the solid phase so produced cannot be characterised by a repeating unit cell arrangement and the matrix is amorphous. Amorphous materials possess no long-range structural order and can be thought of as having the structure of a liquid but the viscosity of a solid.

It is worth noting before commencement of further discussion that the amorphous state is widely studied and discussed yet is generally poorly understood (certainly in quantitative terms). Many of the concepts and terms used in defining and describing amorphous matrices are derived from the polymer field (i.e. large molecular weight compounds that cannot crystallise fully) yet are used to describe small molecular weight pharmaceutical molecules (that usually can crystallise, often to more than one form).

Conceptually, at least, it is possible to envisage a completely and totally amorphous matrix where not even two molecules are in any sort of structural alignment. Such a state would be the most disordered arrangement possible and so would possess the greatest excess enthalpy over the stable crystalline form. In practice such total randomness is extremely unlikely to be achieved, so most models of the amorphous state consider the total matrix to comprise a series of 'microstates', each with a certain degree of short-range structural order. The method of preparation (freeze-drying, spray-drying and so on) is also likely to impart a short-range structure, leading to amorphous materials with reproducibly different macroscopic properties (and so to the notion of polyamorphism – a concept returned to below).

Molecules in any solid phase are never at rest, unless in a perfect crystal at absolute zero, and so movement of molecules (molecular mobility) with time may occur. In a crystalline material molecules may vibrate but rotation and translation are unlikely because of the strong intermolecular interactions forming the crystal lattice energy. In an amorphous material all three modes of molecular movement are permissible, with the result that the structural arrangement of molecules in an amorphous matrix will change with time. Thermodynamics requires that when rearrangement occurs, it happens in a way that reduces excess enthalpy. The result is that amorphous materials will gradually restructure, with short-range order appearing first and, eventually, long-range (crystalline) order. These processes are described as *relaxation* and *crystallisation*. Understanding the rate at which molecular movement occurs is central to understanding (and so controlling) the stability profile of amorphous pharmaceuticals.

The formation of amorphous matrices can be considered from a thermodynamic perspective, starting from the molten state in which the molecules of

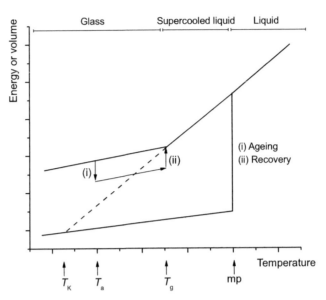

Figure 8.1 Schematic representation of the formation of a glass, showing the glass transition temperature, the Kauzmann temperature, ageing and recovery.

a material are free to move and there is no underlying molecular alignment. If cooled slowly, at the melting temperature of the stable crystalline form the material will crystallise, characterised by a sudden reduction in volume or enthalpy (Figure 8.1). If, as noted in Chapter 7, a perfectly ordered crystal structure is formed, then the energy and volume are minimised and the stable form is obtained. As was also discussed in Chapter 7, if the rate of cooling is sufficiently fast then crystallisation to the stable form may not have time to occur and a supercooled liquid will form. It was assumed earlier that at some point crystallisation would eventually occur to a metastable form.

What would happen if a metastable form could not form (if, for instance, the rate of cooling is too rapid)? Then the volume and enthalpy would continue to decrease along the supercooled liquid line, as shown in Figure 8.1. Although formally the supercooled liquid state is not at a position of thermodynamic equilibrium, the time scales over which molecular movement can occur during this phase are short and so any change in temperature will result in an essentially immediate change in structural and thermodynamic properties. The supercooled liquid is considered to be in *structural equilibrium*.

If the properties of the supercooled liquid were to keep changing linearly with reducing temperature, at some point intersection with the line for the perfect crystal would occur (represented by the dotted extrapolation in Figure 8.1); the temperature of intersection is denoted the Kauzmann temperature (T_K). T_K represents a paradox because, as noted earlier, no arrangement of molecules is possible that results in a smaller volume or enthalpy than the perfect crystal. Some physical transformation must occur before the Kauzmann temperature is reached. That transformation is the formation of a *glass* and it will occur at the *glass transition temperature* (T_g), also shown in Figure 8.1.

Summary box 8.1

- Amorphous means lack of form.

- Amorphous materials have no long-range structural order and excess energy over the crystalline state. This means one barrier to dissolution is removed, leading to rapid dissolution, higher solubility and greater bioavailability.

- Although not easy to define, amorphous materials can be thought of as having the structure of a liquid but the viscosity of a solid.

- Amorphous materials form by rapid cooling. The molecules do not have time to form a crystal and instead exist as a supercooled liquid.

- The glass transition prevents the energy and volume of a supercooled liquid becoming lower than that of a perfect crystal.

Study question 8.1 What physical changes in the material might cause this discontinuity at the glass transition temperature?

It was noted above that the concept of the supercooled liquid being in structural equilibrium depended upon the rate of molecular mobility being reasonably fast, such that structural rearrangement occurred in response to a change in temperature in an essentially instantaneous timeframe. Below the glass transition temperature the rate of molecular mobility is reduced to the point where structural rearrangement does not occur in response to a change in temperature. Effectively, below the glass transition temperature the molecular structure can be considered 'frozen' and the material is a glass (note that this definition does not preclude the movement of molecules over extended time periods – see the concept of relaxation below – just over short time periods). A slightly easier conceptualisation is that an amorphous matrix has the structural arrangement of a liquid but the viscosity of a solid.[1]

---

[1] One consequence of the random structure is that amorphous materials are often transparent, since incident light will not be diffracted in any particular direction, a handy quality of the glass used in window panes.

The discontinuity in molecular mobility leads to two further descriptions of amorphous materials. Above T_g amorphous materials are termed *rubbery* and below T_g they are termed *brittle* (or glassy). These terms originate from the polymer field and are descriptive of the macroscopic physical state of polymeric materials (both are easily demonstrated by freezing an elastic band in liquid nitrogen – brittle – and then allowing it to thaw and pass through its glass transition – rubbery). They are not necessarily descriptive of the macroscopic physical state of small molecular weight organic pharmaceuticals, however, where it can be difficult to imagine a powder as rubbery. Nevertheless, the terms have stuck and the preformulation scientist should be familiar with their meaning.

Study question 8.2   Given the above discussion, what do you think about the concept of polyamorphism?

Summary box 8.2

- As the volume of a supercooled liquid reduces, the viscosity must increase, since the same number of molecules are occupying a gradually reducing volume.

- Above the glass transition temperature, a change in temperature occurs with an essentially immediate change in structure – the matrix is described as rubbery.

- Below the glass transition temperature, a change in temperature does not occur with an immediate change in structure – the matrix is described as brittle or glassy.

8.3 Ageing of amorphous materials

The extrapolated line in Figure 8.1 represents the equilibrium glassy state (in effect, the structural arrangement) that the system would have attained had it not formed a glass. It is often considered a 'virtual' state, because although it can be conceived and defined, it does not actually exist, although it is the structural arrangement to which a real glass will *tend* if it is held at any particular temperature below T_g. The temperature at which the glass is held isothermally is known as the annealing temperature, T_a, and the approach from the glassy state to the equilibrium glassy state is known as *relaxation* (represented by process (i) in Figure 8.1).

Relaxation, as noted earlier, is simply a macroscopic manifestation of the movement of molecules. Two modes of relaxation are usually defined:

structural relaxation and Johari–Goldstein relaxation. Structural relaxation (termed α-relaxation) describes whole molecule movements (such as diffusional motion and viscous flow) while Johari–Goldstein relaxation (termed β-relaxation) is usually taken to represent intramolecular motion (such as rotation of a side-chain). In the context of pharmaceuticals, α-relaxation events are usually considered to be more important than β-relaxation events.

Since it is in a nonequilibrium state, any glass is thermodynamically unstable and its physical properties will eventually change, via relaxation, partial crystallisation or complete crystallisation, as the molecules move and align and the system loses excess enthalpy. The rate of molecular mobility will increase with a rise in temperature and/or the addition of a plasticiser. (Plasticisers are small molecules, often water or organic vapours, that can easily penetrate the spaces between the molecules in an amorphous matrix, altering its physical and thermodynamic properties. In effect, the plasticiser lowers the T_g of the material. When sufficient plasticiser has been added to reduce T_g below storage temperature, the material will recrystallise.) Water is usually a plasticiser for amorphous drug substances (unless they are very hygrophobic), which is why control of RH is often a critical factor in determining physical stability, but many solvents (such as ethanol) act as plasticisers also.

A simple (but quantitative) view of relaxation is that it is the sum of all the changes in molecular configuration that must occur in the amorphous matrix for it to change from the glass to the equilibrium glassy state at a given annealing temperature (T_a). The decay function that characterises relaxation can be expressed as

$$\Phi_{(t, T_a)} = \sum_n a_i \exp\left[-\frac{t}{\tau_i(T_a)}\right] \tag{8.1}$$

where $\Phi_{(t, T_a)}$ is the decay function at time t and annealing temperature T_a, n is the total number of states that are changing independently and a_i is a weighting factor for each state i with an exponential decay constant τ_i.

The number of molecules is obviously large, and so a simpler mathematical interpretation is to consider the matrix as comprising a number of microstates (n); each microstate may contain many molecules but they will all be relaxing at the same rate. Relaxation may then be approximated as the sum of all the changes in configuration of the microstates, usually expressed by the Kohlrausch–Williams–Watts (KWW) equation (Williams and Watts, 1970):

$$\Phi_{(t, T_a)} = \exp\left[-\left(\frac{t}{\tau(T_a)}\right)^\beta\right] \tag{8.2}$$

where τ represents the relaxation time and β is known as the stretch power. Although Equation (8.2) is purely empirical, some physical meaning has been ascribed to its parameters (Phillips, 1994). The stretch power β (which usually has a value between 0 and 1) can be taken to reflect the distribution of independently relaxing states; when β is small, the distribution of the microstates is wide (i.e. there would be a significant number of microstates with widely differing relaxation times) and when β approaches unity the distribution of the microstates is very narrow. In order to describe the dynamics of relaxation in amorphous materials both constants (β and τ) are required. While τ is the time constant and so essentially an indicator of the rate of relaxation, in practice the power term τ^β is found to be a more reliable parameter (Kawakami and Pikal, 2005).

Summary box 8.3

- Molecules in a glassy amorphous matrix will move over extended time periods – relaxation.

- Movement can be whole molecule or intramolecule.

- Relaxation will result in a loss in excess enthalpy and progression of the material towards the supercooled liquid line and, eventually, a crystalline state.

- The physicochemical properties of the material will change as relaxation occurs.

8.4   Characterisation of amorphous materials

Unlike the case of polymorphic materials, there is no need for (nor is it possible to perform) structural characterisation of amorphous matrices (there is no structure). Confirmation that a material is amorphous can be achieved with XRPD. In this case, no specific peaks should be seen; rather, a broad diffraction pattern, known as a halo, is the defining characteristic (Figure 8.2). If a material partially crystallises, peaks will be seen superimposed on to the amorphous halo – the intensity change of any characteristic peak with time can be used as a quantitative indicator of crystallisation rate.

DSC is a particularly useful tool for studying the properties of amorphous materials because, as is apparent from Figure 8.1, many of the changes that occur with time result in a change in enthalpy. The primary indicator of the presence of an amorphous material is a step change in the baseline at T_g. The

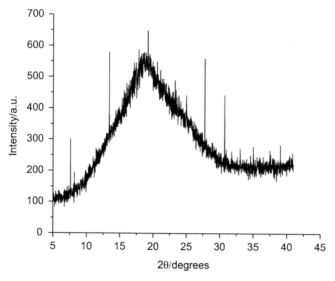

Figure 8.2 XRPD 'halo' seen for amorphous powders.

step change arises because the heat capacity of an amorphous material will be higher above the glass transition temperature as it becomes rubbery (a consequence of the dramatic increase in molecular mobility) – in simple terms, the molecules have more modes in which to convert the energy supplied to them to movement and hence it takes more energy to raise the sample in temperature. Typical DSC data for a glass transition are shown in Figure 8.3. The

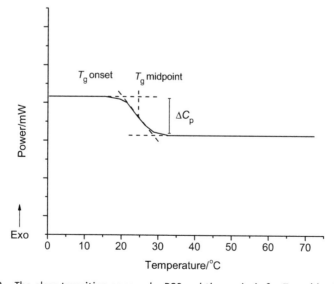

Figure 8.3 The glass transition as seen by DSC and the analysis for T_g and heat capacity.

glass transition temperature can be defined in numerous ways, typically as a deviation from baseline or as an extrapolated mid-point.

The glass transition is a kinetic transformation (involving movement of molecules) and so shows a dependence upon heating rate. In addition, the temperature range over which the transition occurs can be significant.

8.4.1   Measurement of ageing

DSC offers a rapid method to measure the relaxation of amorphous matrices at a particular annealing temperature. The method does not measure the loss in enthalpy as the sample relaxes, but rather is predicated on the fact that when an aged glass is heated, at the glass transition the material will revert back to the equilibrium supercooled liquid state. This occurs because above the glass transition the molecular mobility of the glass is sufficiently high that there will be a change in structural alignment in response to a change in temperature. Since the supercooled liquid line is the equilibrium state, once the molecules have sufficient mobility to reorient themselves, it is this state to which they will tend. The process is represented by line (ii) in Figure 8.1 and the energy is called the enthalpy of recovery ($\Delta_{rec}H$). The enthalpy is assumed to be equal and opposite to the enthalpy of relaxation ($\Delta_r H$).

Study question 8.3   How will the magnitude of the enthalpy of recovery change with annealing time? What does this mean for the sensitivity of the DSC measurement with annealing time?

Relaxation is described by an exponential decay function and so some method must be found to convert the enthalpy of recovery data (which would otherwise be described by a growth function). To do this, the maximum enthalpy of recovery ($\Delta_{rec}H_\infty$) is needed, such that

$$\Phi_{(t, T_a)} = 1 - \frac{\Delta_{rec}H_t}{\Delta_{rec}H_\infty} \tag{8.3}$$

The value of $\Delta_{rec}H_\infty$ is determined from heat capacity change (ΔC_p) at the glass transition:

$$\Delta_{rec}H_\infty = (T_g - T_a)\Delta C_p \tag{8.4}$$

Thus, a plot of $\Phi_{(t, T_a)}$ as a function of time can be constructed (Figure 8.4); fitting these data to the KWW equation (by least squares minimisation) allows the derivation of τ and β.

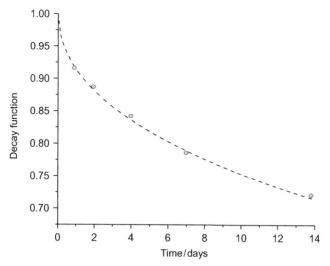

Figure 8.4   Decay function data versus annealing time and the fit to the KWW equation (dotted line).

The rate of decay is affected by both the time constant and temperature. An increase in either causes faster decay, although the time constant is the dominant factor (Figures 8.5 and 8.6).

A second observation is that the DSC will measure the heats associated with both the glass transition and the recovery simultaneously (Figure 8.7). This can complicate quantitative data interpretation.

As before, changing the DSC methodology allows quantitative interpretation of the data. One method is to stop heating the sample once baseline

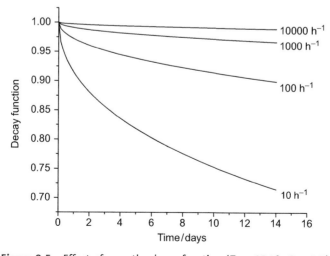

Figure 8.5   Effect of τ on the decay function ($T_a = 25\,°C$, $\beta = 0.5$).

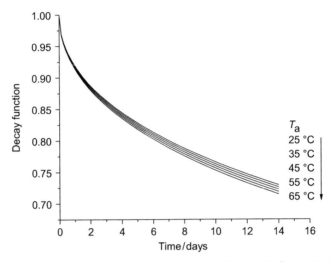

Figure 8.6   Effect of T_a on the decay function ($\tau = 10 \text{ h}^{-1}$, $\beta = 0.5$).

is attained after the enthalpy recovery endotherm. The sample, which at this point is now effectively un-aged, can be cooled to below T_g and reheated. There should be no enthalpy of recovery on the second heating run (unless the material relaxes extremely fast) and so only a step change corresponding to the glass transition should be seen. Subtraction of the second heating data from the first will yield the enthalpy recovery endotherm.

Another option is to use modulated temperature DSC (MTDSC). As noted in Chapter 1, MTDSC separates the power data into two components,

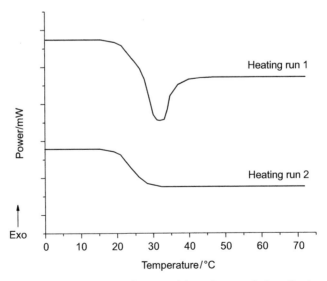

Figure 8.7   DSC thermal trace for the glass transition of an aged glass (top) and the second heat (bottom).

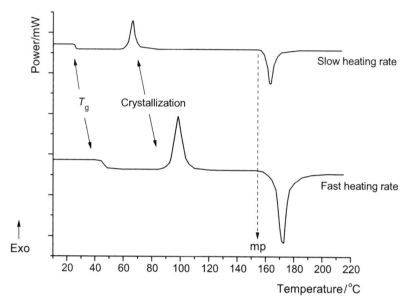

Figure 8.8 Schematic DSC thermal traces for an amorphous material that crystallises and melts, showing the effect of the scan rate.

one representing reversible events and the other nonreversible events. Since the glass transition is reversible but the enthalpy of recovery is not, analysis of an aged glass with MTDSC will allow separation of the events.

For some samples, at temperatures above the glass transition it is possible that no other events may be seen. (This seems counterintuitive because the sample is a solid and should melt. However, the system already contains excess enthalpy over the crystalline form and so the phase transformation to a liquid can proceed with no additional change in enthalpy, just a change in heat capacity.) For other samples the sample may crystallise (exothermic) and the crystal form may melt (endothermic). In this case, the scan rate will affect the DSC thermal curve obtained, with the melting endotherm appearing at the same temperature and every other transition moving to a higher temperature with increasing scan rate (Figure 8.8).

Summary box 8.4

- XRPD and DSC can be used to confirm the presence of amorphous material.

- The diffractogram of an amorphous matrix will be a 'halo'.

- The rate of relaxation can be determined from DSC by measuring the enthalpy recovery as a function of annealing time.

8.5   Processing and formation of amorphous material

Any processing step that imparts a force to a sample or has the capacity to dissolve and condense molecules can potentially lead to formation of an amorphous matrix. Some processing techniques (such as spray-drying, freeze-drying or quench-cooling) will produce almost entirely amorphous matrices while others, such as compaction or milling, result in partially amorphous materials.

8.5.1   Spray-drying

A solution or suspension of the drug substance is forced through an atomising nozzle with heated, pressurised air, producing aerosolised droplets, which are dried in a vortex of air before being separated and collected. Commercially, spray-drying is an efficient process, both because the drying time of each droplet is short (1–2 s) and because it can operate on a continuous feed. Since the particles are produced from droplets, spray-dried material tends to be perfectly spherical, improving flow properties (Figure 8.9). If additional excipients are included in the initial solution or suspension, then spray-drying also offers an excellent method of mixing.

8.5.2   Freeze-drying

The principle of freeze-drying is based on the phase diagram of water (Figure 8.10) and exploits the fact that water has a triple point (it coexists as

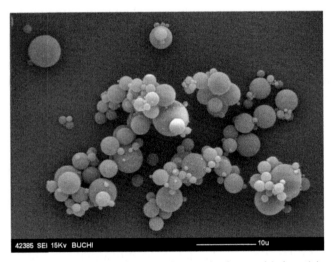

Figure 8.9   A scanning electron micrograph of spray-dried particles.

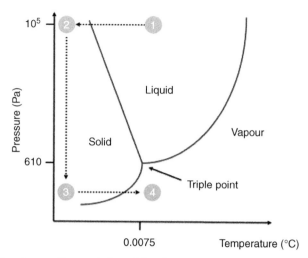

Figure 8.10 The stages of freeze-drying with reference to the phase diagram of water (see the main text for an explanation of the processes at points 1 to 4).

a solid, liquid and gas) at 0.0075 °C and 610 Pa. At pressures below 610 Pa, heating solid water will result in a phase transformation directly to a gas (sublimation).

The material to be dried is loaded into the freeze-drier at room temperature and pressure (point 1 on Figure 8.10). The temperature is reduced and the water freezes (point 2). The frozen sample is then placed under vacuum (point 3) before the temperature is increased (point 4). The water is removed by sublimation, assuming that the pressure is always below 610 Pa. Freeze-dried material tends to be very porous because the process of freezing first produces ice crystals, forming a matrix in which concentrated drug solution pockets are formed, leading to precipitation, and so has a large volume, low density and high surface area (Figure 8.11), leading to very rapid dissolution. This is why some drugs for IV administration are formulated as freeze-dried powders. Being in a solid state confers stability upon storage, but they can be rapidly reconstituted with an aqueous solvent immediately prior to injection, the rapid rate of dissolution ensuring no particles are present.

8.5.3 Quench-cooling

The drug substance is melted (typically on a piece of aluminium foil or a watch glass) and then submersed in a cold medium (such as liquid nitrogen). The rapid cooling rate usually (but not always – griseofulvin is an example of a drug that is almost impossible to make amorphous) results in glass

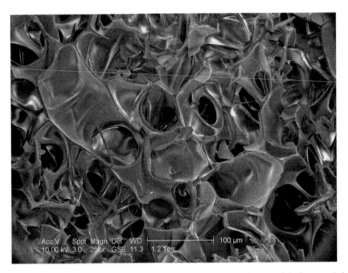

Figure 8.11 A scanning electron micrograph of a freeze-dried material.

formation. The method has the advantages of being quick, cheap and effective, but does not work in cases where the drug substance degrades upon melting nor where crystallisation is extremely rapid.

8.5.4 Milling

Milling is a process used to reduce the particle size of a powder. Pharmaceutically, typical methods include ball-milling (the powder is placed in a jar along with ceramic balls, the mass of the balls being typically tenfold greater than the mass of the powder and the whole is rotated at speeds up to 300 rpm) or air-jet milling (two streams of particles are fired into one another using high-pressure air and impaction causes a reduction in size). In either case, the force applied to reduce particle size is applied at the particle surface. Initially the powder will fracture but there will come a point at which no further reduction in size will occur and any energy supplied is instead used to disrupt crystal structure, leading to the formation of amorphous regions. Although the amount of amorphous material formed in this way might be low (1–5% w/w) it will reside at the surface of the material and so has the potential to dominate the macroscopic properties of the powder. The data in Table 8.1, showing surface energies measured using inverse-phase gas chromatography, illustrate this effect.

Study question 8.4   How might the amorphous content of milled lactose affect its processability? Why do you think the surface energy of milled lactose is higher than that of the spray-dried sample?

Table 8.1 Surface energy of various lactose samples determined with inverse phase chromatography (data from Newell *et al.* (2001)).

Sample	Dispersive energy (mJ m^{-2})
Amorphous lactose (spray-dried)	37.1 (2.3)
Crystalline lactose monohydrate	31.2 (1.1)
Milled lactose (ca. 1% w/w amorphous)	41.6 (1.4)
Mixture (blend of crystal/amorphous)	31.5 (0.4)

8.5.5 Compaction

The high shear forces experienced by a powder bed during compression in a tablet press may well disrupt the crystal structure at the surface of a material in the same manner as described above for milled material. A further consideration is that amorphous material will compress differently to a crystalline material, since the matrix is less dense and the component molecules have greater mobility. Thus an amorphous or partially amorphous material would probably compact to a smaller volume and result in a harder tablet. An example of this effect is discussed in Chapter 11.

Summary box 8.5

- Any process that exerts a force on a sample has the potential to form amorphous material.

- Processes such as spray-drying, freeze-drying and quench-cooling will make samples completely amorphous.

- Processes such as milling and compaction can make samples partially amorphous.

- The percent amorphous content may be low (1–5% w/w) but its influence on powder flow in particular can be dramatic.

8.6 Amorphous content quantification

In cases where a material has been spray-dried, freeze-dried or quench-cooled it is likely that the entire sample has been rendered amorphous. Thus, quantification of the amorphous content is not the issue; an XRPD halo will confirm loss of crystallinity. In cases where a material has been processed and as a result is rendered partially amorphous, knowledge of the percent amorphous content can be very important, given the different physicochemical properties of amorphous materials discussed above and their relationship to processability.

Giron *et al.* (2007) note that analytical techniques for quantification of polyphasic (including partially amorphous) pharmaceuticals should offer both sensitivity and selectivity. The 'gold standard' for structural determination is of course XRPD, but a drawback of the use of XRPD for the study of partially amorphous materials is that it is sensitive to the presence of crystalline structure, rather than the lack of it, which means it can be difficult to use it to quantify amorphous contents below 5% w/w, although it is possible in some cases (Chen, Bates and Morris, 2001). Many analytical techniques have been shown to be capable of detecting amorphous contents to 1% w/w or better, including inverse-phase gas chromatography (IGC), FT-infrared spectroscopy (FTIR), near-infrared (NIR) spectroscopy and solid-state NMR. However, the focus in the current context is on rapid estimation of amorphous content during preformulation and so the best options at this stage are DSC and/or DVS. In either case, a calibration line can only be prepared if standards of known amorphous content are available or can be prepared.

8.6.1   Calibration standards

A series of standard samples of the same chemical composition and polymorph (if appropriate) as the analyte are required but of known amorphous content.

Study question 8.5   How might you prepare a series of standard samples of known amorphous content?

There are a number of factors to consider when preparing calibration standards. Firstly, assuming the standards have been prepared as a powder blend, mixing and sampling is critical in ensuring consistent material. In a study of partially amorphous reference materials for solution calorimetry Hogan and Buckton (2000) showed a better linearity of response if the amorphous fraction of the calibration standard was weighed directly into the calorimetric chamber than if a large sample was blended and samples taken from it. Partly, this is because amorphous powders tend to be both highly cohesive and adhesive (sticking to themselves or the walls of any container in which they are blended).

Secondly, the physical form of the sample is important. The amorphous phase cannot be polymorphic and neither can it comprise stoichiometric hydrates or solvates. Thus, if the sample has polymorphic forms it is critical that the form used for preparation of the calibration standards is the same as that of the sample to be analysed, since it will act as a seed. If a metastable form is used, then care is required to ensure that the 100% amorphous reference material crystallises to the correct form (since no crystal seed is present). If the sample is a hydrate or solvate then the appropriate plasticising vapour should be used to ensure that the hydrate or solvate is formed upon crystallisation (for instance, ethanol should be the plasticising vapour if the sample is an ethanolate).

The situation is different for anomeric sugars. Ramos, Gaisford and Buckton (2005) demonstrated the effect of preparing calibration lines for α and β lactose and showed they were different (the enthalpies of crystallisation of the two forms being significantly different).

Study question 8.6 Why is the case of anomeric materials different from polymorphic materials?

Where samples are milled powders, each particle will comprise a crystalline core surrounded by a corona of amorphous material. Not only will milled powders thus wet differently, because the amorphous regions are located at the surface, but the amorphous fraction will always be in intimate contact with a crystalline seed, which will probably result in a faster rate of crystallisation, noted by Brodka-Pfeiffer *et al.* (2003) in a study of salbutamol sulphate. The issue of seed material also affects the calibration standards. Assuming all particles are the same size and are uniformly blended, then at amorphous contents up to 50% w/w each amorphous particle should be in contact with a crystalline particle. At higher amorphous contents the number of amorphous particles will always exceed that of the crystalline particles and in the limit no crystal particles are present. This can typically mean that high amorphous content samples either do not crystallise fully or may crystallise to a different polymorph or hydrate, meaning that linearity in the calibration line can be lost (an example is given in Figure 8.12).

Finally, the particle size of milled samples is usually small so the surface area is large. If the crystalline material used to prepare the calibration standards has a larger particle size, the heat of wetting will be reduced. If possible, the best option is to prepare the crystalline reference material by 'conditioning' a sample of milled material (i.e. storing it under a suitable plasticising vapour to crystallise any amorphous fraction) to ensure the surface areas are identical.

8.6.2 DSC for amorphous content quantification

Two events might occur during heating in a DSC that are associated with the presence of amorphous material: a step-change in the baseline at the glass

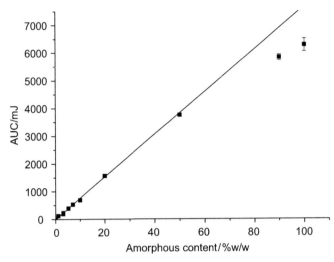

Figure 8.12   A calibration line for the amorphous content in salbutamol sulphate showing deviation from linearity as the amorphous content approaches 100% (redrawn from Gaisford and O'Neill, Copyright 2011, with permission from Elsevier).

transition and an exothermic enthalpy of crystallisation. Either can be used to quantify amorphous content.

The glass transition is the obvious parameter upon which to base an assay, because amorphous materials invariably exhibit one. Moreover, the glass transition temperature is usually reasonably low, reducing exposure of the sample to unnecessary thermal stresses. However, it is sometimes difficult to measure the step height accurately, for two reasons. Firstly, the heat capacity change at T_g can be small and, secondly, as discussed above, the enthalpy of recovery is recorded at the same temperature. In practice, this means that the glass transition is often masked by the enthalpy of recovery and the raw data appears as a peak when a step-change is expected (it can be difficult even to recognise that a glass transition has occurred in these circumstances – one characteristic sign is that there is a step-change in the baseline before and after the peak – let alone measure the step height).

One methodology that allows the separation of these events is to use MTDSC. As discussed in Chapter 1, MTDSC allows the deconvolution of the raw power data into two components (reversing and nonreversing). The glass transition is a reversible event, while the enthalpy of recovery is a nonreversible event, so they will appear in different components.

Reversing and nonreversing data for a sample of trehalose passing through a glass transition are shown in Figure 8.13. The total data (i.e. that recorded with standard DSC) show a complex pattern of events, corresponding to loss of water, the glass transition and the enthalpy of recovery. Of these

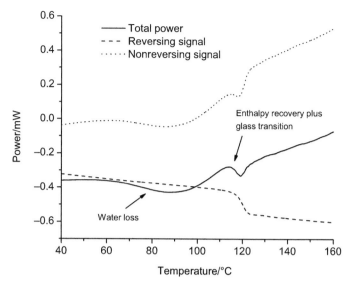

Figure 8.13 Separation of MTDSC data into reversing and nonreversing signals for a sample of trehalose.

events, only the glass transition is reversible and so it also appears in the reversing signal, the remaining events being seen in the nonreversing signal. It is clearly easier to quantitate the glass transition data from the reversing signal. Guinot and Leveiller (1999) and Saklatvala, Royall and Craig (1999) give examples of the use of MTDSC to quantify small amorphous contents in pharmaceuticals.

An alternative option is to use fast heating rates (up to 750 °C min^{-1}). In this case, as discussed in Chapter 7, the magnitude of the step-change will increase with the heating rate. Saunders *et al.* (2004) demonstrated how fast-scan DSC increased the height of the step-change for amorphous lactose from ca. 1 W g^{-1} at 100 °C min^{-1} to 3 W g^{-1} at 250 °C min^{-1} and to 10 W g^{-1} at 500 °C min^{-1}. They also showed that the magnitude of the step-change increased linearly with amorphous content, quoting a limit of quantification of 1.89% w/w and a theoretical limit of detection of 0.57% w/w. A similar study by Lappalainen, Pitkänen and Harjunen (2006) on sucrose showed limits of detection and quantification of 0.062 and 0.207% w/w respectively.

8.6.3 DVS for amorphous content quantification

The principle here is that crystalline material will adsorb water while amorphous material both adsorbs and absorbs water. As water is absorbed, the

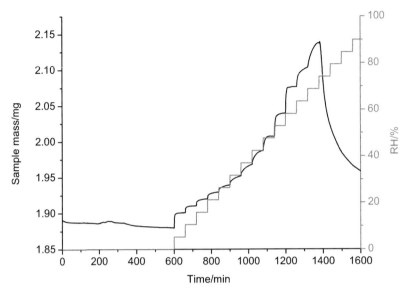

Figure 8.14   DVS data for a sample of amorphous salbutamol sulphate, showing water absorption and then crystallisation at a critical RH of 75%.

amorphous matrix become plasticised and so the rate of mobility of the constituent molecules increases. Eventually, at a critical RH (cRH) the amorphous material recrystallises, a processes that occurs with a concomitant expulsion of the absorbed water. These events are manifest in the DVS data as an increase in mass with RH until cRH, where a sudden fall in mass is seen. Figure 8.14 shows DVS data for amorphous salbutamol sulphate, where the increase in sample mass with increasing RH is seen until 75% RH, where there is a loss in mass (as the sample crystallises).

Recording DVS data in the manner shown in Figure 8.14 allows positive identification of the presence of amorphous material (because of the mass loss at cRH) but it is not easy to use the data as the basis of an assay because of the difficulty in isolating a quantitative parameter that varies with amorphous content. One approach is to consider the difference in water (or solvent) absorption by the crystalline and amorphous regions. Mackin *et al.* (2002) proposed a method. The sample is equilibrated under 30% RH (or relative vapour pressure, RVP, in their specific example), then 85% RH (or any RH above cRH), then 30% RH. During the initial period of equilibration at 30% RH the crystalline material adsorbs water and the amorphous material both adsorbs and absorbs water. Raising the RH above cRH causes crystallisation of the amorphous material. During the second period of equilibration at 30% RH only adsorption of water to crystalline material can occur. The difference in water uptake between the two periods at 30% RH is proportional to amorphous content. Vollenbroek *et al.* (2010) proposed an alternative method based on fitting the moisture adsorption isotherms to a Brunauer, Emmett and Teller (BET) model.

Summary box 8.6

- Assays for quantifying percent amorphous content are essential in cases where the presence of amorphous material is known to affect drug substance properties or drug product performance.

- All assays require preparation of a series of standards of known amorphous content.

- Standards are made by blending mass ratios of amorphous and crystalline material; it must be recognised that the behaviour of such material may be significantly different from the sample to be tested.

- Many analytical techniques can be used to quantify amorphous contents but during preformulation DSC and DVS are quickest and easiest.

- DSC assays are based on either the change in step height of the glass transition or the heat of crystallisation.

- DVS assays are based on changes in water uptake.

8.7 Summary

Amorphous material has no long-range (crystalline) structure and is a high-energy material. This leads to rapid dissolution, higher bioavailability and changes in a range of other physicochemical properties. Formulation in an amorphous form is thus an attractive strategy for overcoming poor aqueous solubility of a crystalline drug substance. However, the amorphous form is unstable relative to a crystalline state, and so ensuring retention of form during storage is critical for drug product performance. The readiness of a drug substance to form an amorphous matrix can be assessed during preformulation by spray-drying, freeze-drying or quench-cooling. Characterisation with XRPD or DSC will confirm the amorphous nature and indicate potential stability. In addition, it should be recognised that milling or compaction can make a substance partially amorphous, an outcome that can materially affect drug substance and product performance. Many methods are available for quantifying amorphous contents, but during preformulation those based on DSC and DVS are quick, near universally applicable and use little sample.

References

Brodka-Pfeiffer, K., Langguth, P., Graβ, P. and Häusler, H. (2003) Influence of mechanical activation on the physical stability of salbutamol sulphate. *Eur. J. Pharm. Biopharm.*, **56**, 393–400.

Chen, X., Bates, S. and Morris, K.R. (2001) Quantifying amorphous content of lactose using parallel beam X-ray powder diffraction and whole pattern fitting. *J. Pharm. Biomed. Anal.*, **26**, 63–72.

Giron, D., Monnier, S., Mutz, M. *et al.* (2007) Comparison of quantitative methods for analysis of polyphasic pharmaceuticals. *J. Therm. Anal. Cal.*, **89**, 729–743.

Guinot, S. and Leveiller, F. (1999) The use of MTDSC to assess the amorphous phase content of a micronised drug substance. *Int. J. Pharm.*, **192**, 63–75.

Hancock, B.C. (1998) Comments on 'An approach to estimate the amorphous content of pharmaceutical powders using calorimetry with no calibration standards'. *Int. J. Pharm.*, **160**, 131–133.

Hogan, S.E. and Buckton, G. (2000) The quantification of small degrees of disorder in lactose using solution calorimetry. *Int. J. Pharm.*, **207**, 57–64.

Kawakami, K. and Pikal, M.J. (2005) Calorimetric investigation of the structural relaxation of amorphous materials: evaluating validity of the methodologies. *J. Pharm. Sci.*, **94**, 948–965.

Lappalainen, M., Pitkänen, I. and Harjunen, K. (2006) Quantification of low levels of amorphous content in sucrose by hyperDSC. *Int. J. Pharm.*, **307**, 150–155.

Mackin, L., Zanon, R., Park, J.M. *et al.* (2002) Quantification of low levels (<10%) of amorphous content in micronised active batches using dynamic vapour sorption and isothermal microcalorimetry. *Int. J. Pharm.*, **231**, 227–236.

Newell, H.E., Buckton, G., Butler, D.A. *et al.* (2001) The use of inverse phase gas chromatography to measure the surface energy of crystalline, amorphous and recently milled lactose. *Pharm. Res.*, **18**, 662–666.

O'Neill, M.A.A. and Gaisford, S. (2011) Application and use of isothermal calorimetry in pharmaceutical development. *Int. J. Pharm.*, **417**, 83–93.

Phillips, J.C. (1994) Microscopic theory of the Kohlraush relaxation constant β_K. *J. Non-Cryst. Solids*, **172–174**, 93–103.

Phillips, E.M. (1997) An approach to estimate the amorphous content of pharmaceutical powders using calorimetry with no calibration standards. *Int. J. Pharm.*, **149**, 267–271.

Ramos, R., Gaisford, S. and Buckton, G. (2005) Calorimetric determination of amorphous content in lactose: a note on the preparation of calibration curves. *Int. J. Pharm.*, **300**, 13–21.

Saklatvala, R., Royall, P.G. and Craig, D.Q.M. (1999) The detection of amorphous material in a nominally crystalline drug using modulated temperature DSC – a case study. *Int. J. Pharm.*, **192**, 55–62.

Saunders, M., Podluii, K., Shergill, S. *et al.* (2004) The potential of high speed DSC (hyperDSC) for the detection and quantification of small amounts of amorphous content in predominantly crystalline samples. *Int. J. Pharm.*, **274**, 35–40.

Vollenbroek, J., Hebbink, G.A., Ziffels, S. and Steckel, H. (2010) Determination of low levels of amorphous content in inhalation grade lactose by moisture sorption isotherms. *Int. J. Pharm.*, **395**, 62–70.

Williams, G. and Watts, D.C. (1970) Non-symmetrical dielectric relaxation behaviour arising from a simple empirical decay function. *Trans. Fara. Soc.*, **66**, 80–85.

Answers to study questions

8.1 There are many theories in the literature, but the following concept provides a workable model. Cooling of the supercooled liquid will result in a concomitant reduction in volume. Since the number of molecules in the sample is unchanging, it follows that the number of molecules per unit volume must increase as the temperature is decreased. In other words, there is an increase in viscosity and, as a result, a corresponding decrease in molecular mobility.

8.2 This reduces to personal choice! The degree of structural order in an amorphous matrix is infinitely variable and will change with time (in the direction of increasing order). While it might be possible to produce amorphous materials by spray-drying and freeze-drying that appear to have reproducibly different properties (and so are polyamorphic), both amorphous matrices will relax to the same final state via an infinite number of intermediate states. This means there can be no transition directly from one defined amorphous state to another (in the way metastable polymorphs might convert, for instance), which, in the view of the authors at least, renders the concept of polyamorphism unsound.

8.3 The enthalpy recovery will increase with an increase in annealing time (because it is the opposite of the energy lost during relaxation). This means that the signal-to-noise ratio of the DSC data improves with increased ageing time (as the measured signal will increase).

8.4 A higher surface energy will result in greater adhesion or cohesion to other particles, leading to clumping and/or sticking to surfaces and, consequently, poorer flow. In addition, if the milled lactose were to be used as a coarse carrier in a dry powder inhaler formulation, there is a danger that the bioavailability of the drug will change because of the increased force of adhesion.

One hypothesis for the higher surface energy of the milled sample compared with the spray-dried sample was alluded to earlier in the text. During spray-drying a droplet of solution is formed – the diameter of the droplet will reduce as the water evaporates. Because water has a high surface tension, it is likely that as condensation occurs, the molecules closest to the surface will become aligned. Thus the surface energy will be reduced, relative to a truly disordered material such as that produced in a mill.

8.5 The only way to prepare samples such that they have a defined amorphous content is to blend nominally 100% amorphous material (typically spray-dried or freeze-dried) with crystalline material. The calibration standards will therefore comprise powder blends where each individual particle is either amorphous or crystalline. Phillips (1997) noted the problems with such an approach and suggested a calculation method as an alternative, although Hancock (1998) subsequently discussed the limitations inherent in the calculation and so it remains the case that powder blends are the only practicable choice.

8.6 Polymorphism is a structural arrangement of whole molecules. Therefore the concept of polymorphism cannot apply to an amorphous matrix. Where a molecule is anomeric, this relates to an intramolecular change in structure and so an anomeric form *is* maintained in the amorphous state. In other words, amorphous α-lactose is a different material from amorphous β-lactose.

9
Stability Assessment

9.1 Introduction

Determining the stability profile of a drug substance or drug product with respect to environmental factors (such as light, temperature and/or humidity) is central to successful development. Advice on stability assessment is provided in ICH Guideline Q1A(R2) (2003) and Q1B (1996), covering both drug substance and drug product, and it is to these guidelines that this chapter is primarily referenced.

Stability studies should focus upon all the attributes of the drug substance that may change during storage and that might influence the quality, safety or efficacy of the drug product. During preformulation, stability assessment is focused almost entirely on the properties of the drug substance itself, rather than the formulated drug product, although initial assessment of likely drug-excipient interactions are useful to inform the design of formulation. The primary attribute would thus be the determination of any chemical routes of degradation and their sensitivity to light, oxygen, humidity, catalysis and/or temperature. If multiple solid-state forms have been identified then the physical stability of the form selected for development should be investigated. Stability in this context may refer to conversion rates to another physical form or to a change in structure (such as the collapse of a freeze-dried matrix). Note that ICH Guidelines define hydrates, solvates and amorphous matrices as distinct physical forms that must be identified and characterised if they impact product performance, bioavailability or stability.

The choice of the degradation limit allowable is arbitrary and will be drug substance-dependent, but usually the lowest acceptable level of potency is 90% of the label claim (Kommanaboyina and Rhodes, 1999). It is also a requirement that the product should look, taste and perform as it did when it was first made.

Essentials of Pharmaceutical Preformulation, First Edition. Simon Gaisford and Mark Saunders.
© 2013 John Wiley & Sons, Ltd. Published 2013 by John Wiley & Sons, Ltd.

Since time is limited during preformulation, stability assessment at this stage usually involves challenging the compound (possibly in combination with likely excipients) by exposure to a range of environmental stresses, most of which are in excess of those that would normally be experienced. ICH Guideline Q1A(R2) (2003) defines four conditions for general stability assessment:

- Long-term storage. The conditions under which the drug substance (and drug product) will be stored. Typically $25 \pm 2\,°C$ and $60 \pm 5\%$ RH, but other values can be specified if justified for a particular drug substance. When submitting data for regulatory approval, at least 12 months data are required.

- Intermediate storage. Typically $30 \pm 2\,°C$ and $65 \pm 5\%$ RH. These conditions may also be selected for long-term storage, in which case no intermediate conditions are needed; 6 months data are required before regulatory submission.

- Accelerated stability. Typically $40 \pm 2\,°C$ and $75 \pm 5\%$ RH; 6 months data are required before regulatory submission.

- Stress testing. Carried out on a single batch of the drug substance as a function of temperature (in $10\,°C$ increments and at temperatures above those used for accelerated stability assessment) and humidity (75% or greater). The effect on oxidation and photolysis should be studied where appropriate, as should the extent of hydrolysis across a range of solution pH.

Table 9.1   ICH storage conditions for general, refrigerated and frozen drug substances and products (ICH Guideline Q1A).

Study	Storage condition	Minimum data required before regulatory submission
General		
Long-term	$25 \pm 2\,°C$ and $60 \pm 5\%$ RH	12 months
	$30 \pm 2\,°C$ and $65 \pm 5\%$ RH	
Intermediate	$30 \pm 2\,°C$ and $65 \pm 5\%$ RH	6 months
Accelerated	$40 \pm 2\,°C$ and $75 \pm 5\%$ RH	6 months
Refrigerated		
Long-term	$5 \pm 3\,°C$	12 months
Accelerated	$25 \pm 2\,°C$ and $60 \pm 5\%$ RH	6 months
Frozen		
Long-term	$-20 \pm 5\,°C$	12 months

Table 9.2 ICH climatic zones and their associated long-term storage conditions.

Climatic zone	Definition	Long-term storage conditions
I	Temperate	21 °C, 45% RH
II	Subtropical and Mediterranean	25 °C, 60% RH
III	Hot and dry	30 °C, 35% RH
IVA	Hot and humid	30 °C, 65% RH
IVB	Hot and very humid	30 °C, 75% RH

Conditions are also specified for refrigerated and frozen substances and products (Table 9.1). Two points are of note here. Firstly, during preformulation development the aim is rapidly to identify any instability and, where possible, to determine a causative mechanism. At this stage, therefore, stress testing is the focus, with the objective of identifying any drug substances with unacceptable stability profiles. Long-term and accelerated studies would be performed later in development. Secondly, environmental conditions vary considerably around the world. Stability data conducted at 25 °C in London may not successfully indicate likely stability in India at 45 °C and high humidity, for instance. ICH defines four climatic zones (I–IV) (Table 9.2), based on the work of Grimm (1986, 1998). The stability conditions defined in Table 9.1 apply only to climatic regions I and II. These are assumed to represent Europe, Japan and the United States and so stability data generated in one of these regions should be mutually acceptable in the other two. Stability conditions for regions III and IV would need to be amended as appropriate.

> *Summary box 9.1*
>
> • Stability testing protocols are defined in ICH Guidelines Q1A(R2) and comprise long-term, intermediate, accelerated and stress conditions.
>
> • Long-term storage can also mean refrigerated or frozen, depending on the stability of the drug substance.
>
> • Storage conditions vary around the world, so four climatic zones are defined.

9.2 Degradation mechanisms

Chemical degradation usually occurs via one of three reaction types:

- Hydrolysis or solvolysis
- Oxidation
- Photolysis

The rates of all reaction types may be increased with a catalyst. In addition, physical changes may occur that affect product performance, such as a change of polymorph or loss/formation of a hydrate form. The primary outcomes of stress testing during preformulation should be to determine:

(1) The potential for hydrolysis (and hygroscopicity of the compound).

(2) The potential for solvolysis.

(3) The potential for oxidation (and the use of anti-oxidants to increase stability).

(4) The potential for photolysis (and determination of any causative wavelengths).

(5) The order of reaction (giving an insight into the mechanism).

(6) The influence of pH on the kinetics and extent of degradation reactions in aqueous solution. Manipulation of pH (within physiologically acceptable limits) is an important tool in minimising degradation.

(7) The temperature dependence of the reaction kinetics.

(8) The influence of relative humidity on reaction kinetics.

(9) The relative stability of polymorphic or pseudopolymorphic forms including ranking of metastable forms and identification of the thermodynamically stable form.

This information should be sufficient in order to make a go/no go decision in respect of each drug substance. Once the underlying degradation profile of a drug substance is understood, dosage form, excipient choice, processing conditions and packaging selection can be made to mitigate any stability issue. The rate of degradation in the solid-state usually varies between polymorphic or pseudopolymorphic forms. Stress testing in solution will identify primary degradation pathways, but solid-state stability testing of the physical form selected for development will be necessary in the medium-term. Long-term and accelerated stability programmes can then be designed and implemented to provide data for regulatory submission. Since physical form characterisation was discussed in Chapters 7 and 8, the text below considers chemical degradation.

9.2.1 Hydrolysis

Hydrolysis reactions involve chemical reaction between at least two species, one of which is water. Although hydrolysis reactions should follow second-order kinetics (see below), because water is present in excess the reaction rate will be experimentally measured as pseudo first-order, being dependent upon the concentration of solute only. Chemical groups that are likely to hydrolyse include esters, β-lactams, amides and some imines. Generally amide bonds hydrolyse more slowly than esters. This means amide drug substances may be more stable than their ester counterparts. For instance, procaine (an ester) hydrolyses during autoclaving while procainamide does not. Similarly, hydrolysis rates are hugely dependent upon chemical structure, reflecting the ease with which nucleophilic attack can occur or the stability of any intermediate structures; procaine has a short shelf life in solution (less than 2 days at 25 °C; Loucas *et al.*, 1981), while lignocaine is remarkably stable with respect to temperature, acid and alkali.

In a review of the literature, Day and Ingold (1941) proposed a standard nomenclature for eight standard mechanisms of hydrolysis. The system uses four characters. The first character is either A or B, for acid or base hydrolysis, and the last character is either 1 or 2, for unimolecular or bimolecular. The middle two characters were originally prime or double prime but were revised by Ingold (1953) to AC or AL, representing scission of either the acyl or alkyl oxygen respectively. The two most common mechanisms are $A_{AC}2$ and $B_{AC}2$, reflecting the fact that the acyl bond is more easily broken than the alkyl bond and that bimolecular mechanisms are favoured over unimolecular mechanisms. Both $A_{AC}2$ and $B_{AC}2$ proceed via formation of a tetrahedral intermediate and it is thought that two molecules of water are involved (tetrahedral intermediates require three molecules to collide and interact, which is unusual, but in this case the two water molecules are already interacting through a hydrogen bond). For instance, in the case of $A_{AC}2$,

$$RC(OH)OR' + 2H_2O \rightarrow RC(OH)_2OR' + H_3O^+$$

The $A_{AL}1$ mechanism also proceeds very rapidly in cases where the R' group can exist as a stable carbonium ion.

Where specific acid or base-catalysed hydrolysis occurs, the stability–pH profile of a drug substance will not be linear. The pH of maximum stability typically falls between 2.5 and 7, which reflects the fact that the hydroxyl ion attacking a carbonyl carbon (of an ester or amide) is a better nucleophile

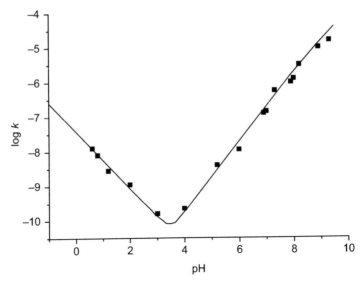

Figure 9.1 V-shaped pH–stability profile for hydrolysis of atropine (redrawn from Connors *et al.* (1986)).

than water attacking a protonated carbonyl. Stability–pH profiles show various behaviours, depending upon the mechanism of reaction:

- V- or U-shaped. Indicative of specific acid catalysis at low pH and specific base catalysis at high pH. Figure 9.1 shows the V-shaped profile for atropine.

- Sigmoidal. Indicative of dissociation of the functional group undergoing nucleophilic attack. Figure 9.2 shows the sigmoidal profile for cycloserine.

- Bell-shaped. Indicative of dissociation of two functional groups. The overall profile is the sum of two mirror-image sigmoidal curves and maximum instability is seen at the point of intersection. Figure 9.3 shows the bell-shaped profile for aztrenam.

The experimentally observed profile may comprise more than one of these behaviours. For instance, the profile for aspirin (Figure 9.4) shows V-shaped behaviour at the extremes of pH, with sigmoidal behaviour in between.

Study question 9.1   It seems sensible to formulate a drug substance at the pH of maximum stability. Why might this be a problem?

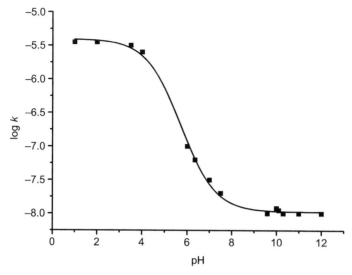

Figure 9.2 Sigmoidal pH-stability profile for hydrolysis of cycloserine in dilute solution (redrawn from Kondrat'eva *et al.* (1971), with kind permission from Springer+Business Media B.V.).

9.2.2 Solvolysis

Solvolysis is the same process as hydrolysis but the reactant is a solvent other than water (for instance, reaction with methanol is methanolysis, reaction with ethanol is ethanolysis, etc.).

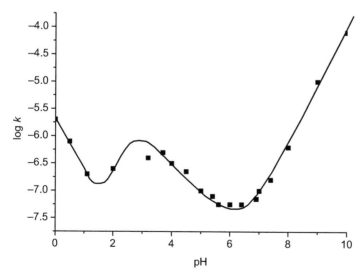

Figure 9.3 Double bell-shaped pH–stability profile for hydrolysis of aztreonam (redrawn from Connors *et al.* (1986)).

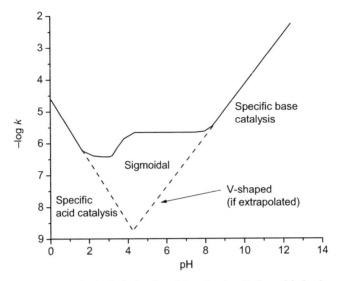

Figure 9.4 pH profile for the hydrolysis of aspirin, showing V-shaped behaviour at extremes of pH and sigmoidal behaviour in between (redrawn from Connors *et al.* (1986)).

Drug substances are generally more stable in organic solvents than in water, although photolysis and oxidation reactions may proceed at a faster rate as free radicals are usually more stable in less polar solvents. Thus, solvent polarity rather than pH is the greatest factor affecting degradation rates. As a general rule if the degradation products are more polar than the parent molecule then addition of a less polar solvent will increase stability. Similarly, where the degradation products are less polar than the parent molecule then addition of a more polar solvent will increase stability. If the drug substance and any degradants are nonpolar, such as the steroids, then there will be no change in stability with polarity and so any (physiologically acceptable) solvent may be added to increase solubility without affecting stability.

9.2.3 Oxidation

Oxidation originally meant reaction of a compound with oxygen to produce an oxide, but is now recognised as any reaction where the oxidation state of the reacting molecule is increased (in other words, it loses at least one electron). By definition therefore the oxidising agent must itself gain at least one electron and is *reduced*, forming a free radical. Oxidation reactions are thus said to involve a *redox* pair. Typically, oxidising agents are compounds that contain elements in high oxidation states (such as H_2O_2, MnO_4^- or $Cr_2O_7^{2-}$) or highly electronegative atoms that can easily accommodate an extra electron (such as O_2, Fe_2 or Cl_2). From a stability perspective, reaction with oxygen is

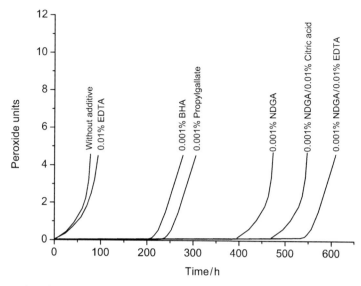

Figure 9.5 The effect of various antioxidants, chelating agents and antioxidant/chelating agent mixtures on the formation of peroxides in oil of terebinth (redrawn from Fryklöf (1954)).

of the greatest interest, since it is present in the environment while other oxi-dising agents would have to have been specifically added to the formulation, although metal ions may be present in solutions to an appreciable degree and these often catalyse oxidative processes. One method to check whether metal ions are catalysing oxidation is to add a metal chelator (such as ethylenedi-amine tetra-acetic acid, ETDA) or antioxidant to the solution; this will bind any metal ions, reducing oxidation of the drug substance (Figure 9.5).

   When a drug substance reacts with atmospheric oxygen spontaneously at room temperature the process is known as auto-oxidation. Molecular structures that tend to oxidise include heterocyclic aromatic rings, those with a hydroxyl group bonded directly to an aromatic ring, conjugated dienes, nitroso and nitrate derivatives and aldehydes. Examples of drugs that are susceptible to auto-oxidation include adrenaline, ascorbic acid, hep-arin, hydrocortisone, morphine, the penicillins, the tetracyclines and some of the oils used in flavouring. One outcome from this is that as well as loss of potency, the drug product may also start to taste unpleasant with time. It is also typical that degradants of oxidation are coloured. Even when the amount of degradation is within tolerance, the degradants may discolour the product.

Study question 9.2   How might oxidative degradation be reduced and, if some oxidation occurs during storage, how might discolouration be masked in the drug product?

9.2.4 Photolysis

Photolysis means reaction of the drug substance or product with light. Demonstration of the photostability of a new drug substance is a requirement of any development programme and is covered by ICH Guideline Q1B (1996). The intrinsic photostability characteristics of a drug substance should be characterised so that exposure to light does not result in an unacceptable change. During preformulation, photostability testing may be performed on a single batch of drug substance under stress conditions. If instability is seen, then confirmatory studies must be performed under standard conditions. Two light sources are specified:

- Any light source that produces an output similar to the D65/ID65 emission standards (these are the outdoor daylight (D65) and indoor daylight (ID65) standards defined in ISO 10977, 1993). Typical examples of such light sources include fluorescent lamps with visible and UV outputs and xenon or metal halide lamps. Significant light emission below 320 nm may be removed with a filter.

- A combination of a cool white fluorescent lamp (producing an output similar to that specified in ISO 10977, 1993) and a near-UV fluorescent lamp with a spectral distribution between 320 and 400 nm and with a maximum output between 350 and 370 nm.

Light is a form of energy, propagated as an electromagnetic wave, and so irradiation of a drug substance with light can potentially initiate several processes. Energy may be absorbed by a drug substance, transferred to other molecules or emitted at a different frequency, resulting in possible degradation and/or an increase in temperature. Light energy may also promote oxidation and hydrolysis, so photostability testing in solution is as important as photostability testing in the solid-state. Since light energy is inversely proportional to wavelength, UV frequencies generally cause more degradation than visible frequencies.

Photolysis can involve multiple mechanisms and reaction pathways and a complex mixture of degradants can be created; the exact composition will be dependent upon the spectral distribution of the light source. Where maximum degradation is seen to occur at a specific frequency of light (the causative wavelength) packaging can be designed to reduce exposure of the drug substance to that wavelength. Plain glass absorbs more than 80% in the 290–320 nm region, while amber glass increases absorption to nearly 95%. It is also possible to add UV blockers or a film coat with an opaque layer.

Where a drug is photolabile, degradation can occur very rapidly (nifedipine, for instance, has a half-life of only a few minutes). One point to note is that rapid degradation at a specific wavelength (photosensitivity) can be a formulation strategy, often used as a targeting strategy for chemotherapy agents.

> **Summary box 9.2**
>
> - The principal causes of chemical degradation are hydrolysis (or solvolysis), oxidation and photolysis.
>
> - Change in physical form may also reduce drug product performance.
>
> - Hydrolysis is often catalysed in acidic or basic conditions. A stability–pH profile is useful in identifying a hydrolysis mechanism.
>
> - Oxidation can lead to change in taste and/or discolouration as well as loss of drug substance.
>
> - Catalysis generally acts to speed hydrolysis and oxidation.

9.3 Reaction kinetics

Most drug products are formulated as solids, and so solid-state reaction kinetics generally govern the shelf life of drug products. Initial formulations, for toxicology or bioavailability studies, however, may well be solutions, in which case solution-state kinetics predominate, or suspensions, in which case pseudo zero-order kinetics often predominate. In either event, understanding the principles of reaction kinetics will facilitate quantitative interpretation of stability data and indicate the potential reaction mechanism.

9.3.1 Solution-phase kinetics

Solution-phase kinetics are based on the concept of *molecularity* (i.e. on reaction stoichiometry). This posits that in solution reactants are free to move and so the reaction rate is proportional to the number of collisions between molecules. Consider the reaction scheme

$$aA + bB \rightleftharpoons cC + dD \tag{9.1}$$

The reaction rate can be defined in terms of the forward reaction (the reacting species):

$$\text{Rate} = k_f \, [A]^a \, [B]^b \tag{9.2}$$

or in terms of the reverse reaction (the products):

$$\text{Rate} = k_r\,[C]^c\,[D]^d \tag{9.3}$$

where the superscripts (a and b or c and d) sum to give the *order* of reaction and k is a constant of proportionality. At equilibrium the rates of the forward and reverse reactions must be equal and so

$$k = \frac{k_f}{k_r} = \frac{[C]^c\,[D]^d}{[A]^a\,[B]^b} \tag{9.4}$$

where k is termed the *rate constant*. Solution-phase reactions are typically zero, first or second order. Noninteger orders may be experimentally measured, but this indicates that degradation is progressing via multiple pathways.

9.3.2 Zero-order reactions

If the order of reaction is zero, then the reaction rate is independent of the concentration of any of the reacting species and is constant:

$$\text{Rate} = \frac{-d[A]}{dt} = k \tag{9.5}$$

When integrated, Equation (9.5) becomes

$$[A]_t = [A]_0 - kt \tag{9.6}$$

where $[A]_t$ is the concentration of reactant A at any time and $[A]_0$ is the initial concentration of reactant A. Hence a plot of concentration versus time is linear with slope $-k$ (Figure 9.6). There are very few truly zero-order reactions (if a reaction is seen to degrade with zero-order kinetics it is usually pseudo zero-order, principally because one of the reacting components is present in great excess – this is the case for suspensions noted earlier, where the solution concentration is kept constant by dissolution of material from the solid phase). Zero-order rate constants have units of concentration time^{-1}.

Two important terms are defined for stability studies: shelf-life and half-life. Shelf-life is the time taken for the concentration of active to fall to a specified percentage of the label claim. Any value can be chosen, but 90% of the label claim ($t_{0.9}$) is a common value. Half-life is the time taken for the

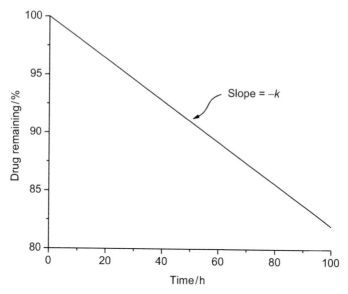

Figure 9.6 Decomposition profile for a drug substance degrading via zero-order kinetics.

concentration of active to fall to 50% of the label claim ($t_{0.5}$). These are easily calculated from the equation of the straight-line plot:

$$t_{0.9} = \frac{[A]_0}{10k} \tag{9.7}$$

$$t_{0.5} = \frac{[A]_0}{2k} \tag{9.8}$$

Study question 9.3 See if you can derive Equations (9.7) and (9.8).

9.3.3 First-order kinetics

First-order kinetics result when the indices sum to 1, which usually means the reaction rate is dependent upon the concentration of one of the reacting species only:

$$\text{Rate} = \frac{-d[A]}{dt} = k[A] \tag{9.9}$$

or, in integrated form,

$$[A]_t = [A]_0 e^{-kt} \tag{9.10}$$

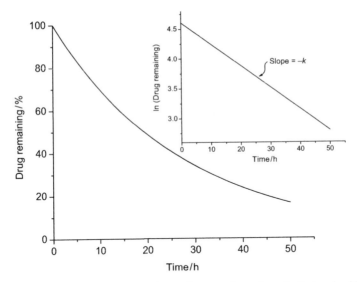

Figure 9.7 Decomposition profile for a drug substance degrading via first-order kinetics and (inset) the linear plot of ln (drug remaining) versus time.

A plot of concentration versus time will show an exponential decay and so the slope cannot be used to calculate the rate constant (Figure 9.7). However, by taking logarithms the following expression is obtained:

$$\ln [A]_t = \ln [A]_0 - kt \tag{9.11}$$

in which case a plot of ln [A] versus time should be linear and of slope $-k$ (the units of a first-order rate constant are time^{-1} since ln values have no units and so the y axis is unitless) (inset graph on Figure 9.7). For first-order degradation the shelf life and half-life are calculated thus:

$$t_{0.9} = \frac{0.1054}{k} \tag{9.12}$$

$$t_{0.5} = \frac{0.693}{k} \tag{9.13}$$

Study question 9.4   See if you can derive Equations (9.12) and (9.13).

9.3.4   Second-order reactions

In this case the reaction indices sum to 2, which can be achieved in two ways:

$$\text{Rate} = \frac{-d[A]}{dt} = 2k [A]^2 \text{ or } 2k [B]^2 \tag{9.14}$$

$$\text{Rate} = \frac{-d[A]}{dt} = k [A][B] \tag{9.15}$$

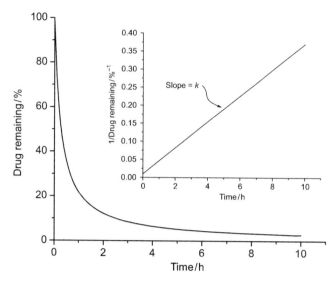

Figure 9.8 Decomposition profile for a drug substance degrading via second-order kinetics and (inset) the linear plot of 1/drug remaining versus time.

or, in integrated forms,[1]

$$\frac{1}{[A]} = \frac{1}{[A]_0} + kt \tag{9.16}$$

$$\frac{[A]}{[B]} = \frac{[A]_0}{[B]_0} e^{([A]_0 - [B]_0)kt} \tag{9.17}$$

In the case of Equation (9.16), a plot of $1/[A]_t$ versus time should be linear (Figure 9.8). In the case of Equation (9.17), where the reaction rate depends upon both reactants, the kinetics are complex to analyse, so it is often the case that the concentration of one of the components is held in excess, which reduces the kinetics to pseudo first order (this is the case for the kinetics of hydrolysis reactions, as noted earlier). The half-life can be calculated from Equation (9.16) as

$$t_{0.5} = \frac{1}{k[A]_0} \tag{9.18}$$

9.3.5 Solid-state kinetics

Unlike reactions progressing in solution, solid-state reactions cannot be described in terms of molecularity because the molecules are much less

---

[1] Note that $[A]_0$ and $[B]_0$ must be different for Equation (9.17) to be valid.

mobile than in solution and so the rate cannot be defined by the frequency of collisions. Instead, solid-state kinetics are based on the fraction (α) of reaction that has occurred as a function of time. In the general case,

$$\text{Rate} = \frac{d[\text{A}]}{dt} = kf(\alpha) \tag{9.19}$$

where $f(\alpha)$ is some function of the progression of reaction. Numerous functions have been described in the literature and functions can be written for specific processes (degradation, crystal growth, etc.).

Determination of the amount of drug substance remaining during solid-state stability trials must be undertaken with care. Firstly, reaction rates are usually appreciably slower than those occurring in solution, which means greater stress conditions must be used or the trial must run over a longer time period. Secondly, determination of concentrations in the solid phase per se is problematic, which means the components must often be dissolved in a solvent prior to (possible) separation by HPLC and quantitative analysis. It must be ensured that no additional degradation occurs following dissolution. Finally, solid–solid phase transitions may be an important factor to consider in terms of loss of potency, and these cannot be identified with any assay that requires solubilisation of the drug substance prior to analysis.

It is also easy to assume that solid materials are free of water or solvent but this is not usually the case. Tablets may contain 2–5% water to facilitate consolidation during compression. This free water may plasticise any amorphous regions, increasing the likelihood of crystallisation and also increasing reaction kinetics. It may also act as an intermediate to facilitate solid-state reactions between the drug substance and any excipients. Also, most 'solid-state' reactions occur primarily within the saturated solutions that will likely form on the surface of any crystals. The rates of reaction in these solutions may be significantly different from those determined from stability trials, because the concentrations are saturated, meaning the activities of the dissolved species will be high. In addition, the ionic strength, pH and polarity of the solutions formed may well differ from expectation. The presence of water may also lead to hydrolysis, although again the rates measured in the solid-state may be significantly slower than determined in solution since the water may not be present in great enough quantity truly to be considered in excess. In this case, the reaction kinetics will change, typically from first or pseudo first order to zero order or, where the main degradation path is intermolecular aminolysis, to second or higher orders.

Study question 9.5　　Why do you think reactions often proceed with zero-order kinetics in the solid-state?

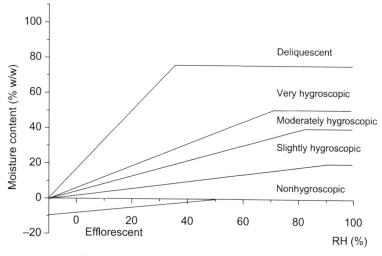

Figure 9.9 Definition of hygroscopicity profiles.

A final factor to consider is the hygroscopicity of the drug substance or excipients. While it might be possible (and desirable) to keep the water content of a drug product low during manufacture, if the product will absorb water during storage any stability issue will not be mitigated. Hygroscopicity can be graded in terms of water uptake as a function of humidity (and is easily measured with DVS) (Figure 9.9). Highly hygroscopic materials can absorb so much water they dissolve themselves. This behaviour, termed *deliquescence*, is most commonly seen with salts (in fact the deliquescence point of salts can be used to calibrate the humidity generators on analytical equipment). For deliquescence to occur the saturated solution formed on the surface of the crystals must have a lower vapour pressure than that of the atmosphere under which it is held.

At the other end of the hygroscopicity scale, some hydrates can lose water of crystallisation to form a lower hydrate or become anhydrous (the process of *efflorescence*). Examples are codeine phosphate (1.5 H_2O to 0.5 H_2O) and sodium carbonate (10 H_2O to 1 H_2O). As water migrates through a drug product to evaporate from the surface, it may well aid diffusion of other compounds, which then precipitate at the surface. This effect leads to 'blooming' of chocolate (crystallised fats on the surface) but can also lead to crystallisation of drug substances from polymer networks, such as hydrogels and oral films.

Between these extremes, materials range in their degree of hygroscopicity. Materials that do not take up any water are described as *nonhygroscopic*, although even these will usually absorb water once the RH increases beyond 95%.

Summary box 9.3

- Solution-phase stability is based on the concept of molecularity.

- Reaction order is an integer – noninteger orders indicate that more than one process is occurring.

- Zero-order rate constant obtained from slope of concentration versus time plot.

- First-order rate constant obtained from slope of ln(concentration) versus time plot.

- Second-order rate constant obtained from slope of 1/(concentration) versus time plot.

- Solid-state reactions are described in terms of the fraction of reaction that has occurred.

9.4   The temperature dependence of reaction kinetics

From experience, reactions tend to happen faster with an increase in temperature. To understand this phenomenon it is necessary to understand the changes in energy that occur as a reaction progresses – thermodynamics. In simple terms, a reaction will progress spontaneously if there is a decrease in Gibb's free energy (ΔG). The Gibb's energy is a function of both the enthalpy (ΔH) and entropy (ΔS) of the system. These are complicated concepts that are themselves the subjects of entire textbooks, but in essence the enthalpy change is a change in heat content and the entropy is a change in order (for spontaneous processes, disorder always increases). The three are related in the well-known equation

$$\Delta G = \Delta H - T\Delta S \tag{9.20}$$

Hence, ΔG is most likely to be negative when the enthalpy change is large and negative (exothermic) and the entropy change is positive (increase in disorder). It is possible for spontaneous reactions to occur with a positive (endothermic) change in enthalpy but only if the change in entropy is very large. Such reactions are termed entropically driven (examples include the dissolution of KCl or mannitol in water, which results in a decrease in solution temperature, the binding of a drug to a biological target, where water molecules are displaced from the binding site or evaporation of volatile solvents), but in practice most spontaneous reactions are exothermic.

Knowledge of the change in enthalpy between the reactants and products determines whether a process is thermodynamically feasible but does not in itself indicate the rate of reaction. Liquid hydrogen and oxygen can be safely mixed at room temperature without interacting, despite the extremely favourable and rather energetic change in enthalpy upon reaction, which suggests there is another factor that governs spontaneity. This is the activation energy (which can be considered as an 'energy barrier' that must be overcome before reaction occurs). The concept posits that reaction between molecules can only occur following collision if their combined energies are greater than the activation energy. This implies that the reaction rate must be a function of both collision frequency and molecular energy. The changes in energy for exothermic and endothermic reactions, and the activation energy are represented diagrammatically in Figure 9.10.

One way of understanding this relationship is to consider a Maxwell–Boltzmann distribution of energies among the molecules in a sample. This states that not all the molecules in a particular sample have the same energy – there is a distribution of energy values and the number of molecules

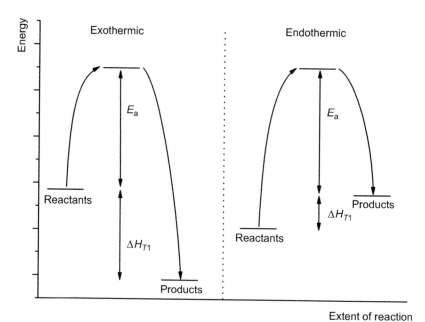

Figure 9.10 Energy diagrams for exothermic and endothermic reactions at a given temperature.

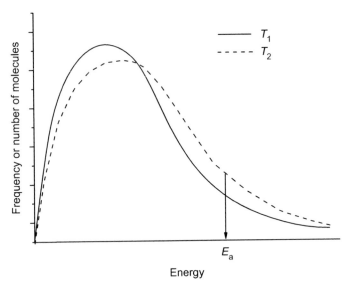

Figure 9.11   Maxwell–Boltzmann distribution of energies at two temperatures ($T_2 > T_1$). The fraction of molecules (area under the curve) possessing energies above E_a will react upon collision.

(or frequency) possessing each particular energy value will differ (Figure 9.11). It can be seen that as the number of molecules possessing a certain energy level falls the energy level increases (the area under the curve gives the fraction of molecules with a certain energy). Only those molecules that have energies above the activation energy will react upon collision. The distribution will change with temperature. In Figure 9.11 it is assumed that T_2 is greater than T_1. Note how the distribution has shifted with an increase in temperature, with the peak maximum being lower but the overall peak being wider. The net result is that the fraction of molecules possessing energies greater than the activation energy (the area under the curve to the right of E_a) has increased. In other words, more molecules will react upon collision at the higher temperature, increasing the observed rate of reaction.

Arrhenius first made empirical observations that reaction rate and temperature were related:

$$\ln k = f\left(\frac{1}{T}\right) \tag{9.21}$$

The relationship was formalised by the work of van't Hoff, leading to the form of the Arrhenius equation used today:

$$k = Ae^{-E_a/RT} \tag{9.22}$$

where A is a constant called the Arrhenius pre-exponential factor, which is essentially related to the fraction of molecules that have the energy to react. Alternatively,

$$\ln k = \ln A - \frac{E_a}{RT} \tag{9.23}$$

van't Hoff also showed the relationship between reaction rate (k) and temperature was given by

$$\ln k = \frac{-\Delta H}{RT} - \frac{\Delta S}{R} \tag{9.24}$$

where R is the universal gas constant ($8.314 \text{ J K}^{-1} \text{ mol}^{-1}$). It follows that a plot of $\ln k$ versus $1/T$ should be a straight line of slope $-E_a / R$. A series of stress tests conducted at elevated temperatures allows determination of a series of rate constants and hence construction of an Arrhenius plot, which can subsequently be extrapolated to storage conditions (Figure 9.12).

Several caveats must be understood when using the Arrhenius plot as the basis of a stability assessment. Firstly, there is assumed to be no change in reaction mechanism over the temperature range of both the experimental measurements and the extrapolation. Secondly, the enthalpy of reaction is assumed to be constant. Consideration of the thermodynamic data presented earlier in Figure 7.5 shows this is not usually valid, although the temperature

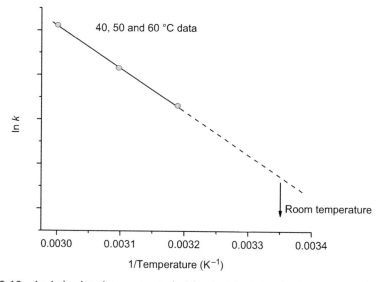

Figure 9.12   An Arrhenius plot constructed with rate data determined at 40, 50 and 60 °C and the extrapolation to room temperature.

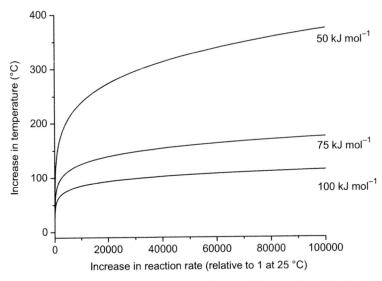

Figure 9.13 Increase in temperature required to increase the rate of reaction (relative to the rate at 25 °C) for reactions with activation energies of 50, 75 and 100 kJ mol^{-1}.

range over which most stability data are recorded is usually narrow enough that curvature in the Arrhenius plot is not apparent. Finally, if the rate of reaction is slow, the increase in temperature required to achieve a measurable extent of degradation within an acceptable time period can be significant and is dependent upon the activation energy. Figure 9.13 shows the temperature rise required to raise the reaction rate relative to the rate at 25 °C for reactions with activation energies of 50, 75 and 100 kJ mol^{-1} (selected values are provided numerically in Table 9.3). For instance, to increase the rate a

Table 9.3 Increase in temperature required to raise the reaction rate relative to the rate at 25 °C as a function of activation energy.

Increase in rate relative to 20 °C	Activation energy (kJ mol^{-1})		
	50	75	100
1	0	0	0
2	10.2	6.8	5.0
4	21.2	13.8	10.2
8	33.0	21.2	15.6
10	37.0	23.7	17.4
100	84.8	51.5	37.0
1000	148.7	84.8	59.3
10 000	238.5	125.1	84.8
100 000	374.3	175.0	114.2

thousandfold for a reaction with an activation energy of 75 kJ mol^{-1} would require an increase in temperature of 85 °C – stressful conditions indeed!

Summary box 9.4

- Reaction rate will vary with temperature.

- An increase in temperature will increase the rate of reaction.

- An Arrhenius plot of high-temperature reaction rates allows extrapolation to long-term storage conditions.

- Temperature rise required to accelerate reactions can be significant.

9.5 Stress testing

9.5.1 Stress testing in solution

It was noted earlier that stress testing is the main focus during preformulation, since it is at this stage that identification of any possible degradation routes is necessary. Of course, likely mechanisms of degradation may also be predicted based on drug class or consideration of functional groups. Whether such degradation will occur during storage or post formulation can be determined (or mitigated) later during development.

An easy starting point is to degrade the drug substance under extreme conditions by refluxing with 5 M HCl and also 5 M NaOH. The resulting concoction can be studied by separating the degradation products with TLC and/or HPLC. Note at this stage that the structures of the degradants will not be known, so the only discriminating assay available will be for the drug substance itself, which is why stability profiles usually plot loss of the drug substance with time. Once the major degradation products have been identified, specific assays may be developed for them.

It is also important to establish early on the identity of the predominant degradation mechanism. As has been noted before, during preformulation the amount of drug substance may be extremely limited and so a scheme that involves the minimum experimentation to distinguish between hydrolysis, oxidation, photolysis and temperature would be beneficial. Wells (1988) suggested such a scheme (Table 9.4). A solution of drug substance is prepared over a range of pH values (indicating whether hydrolysis is occurring and if it is acid or base catalysed). Samples are also stored in the presence or absence of oxygen, in the presence or absence of light and at elevated temperature. Table 9.4 indicates the outcomes that would be seen in the event of

Table 9.4   A scheme to determine the mechanism of degradation. Key: O_2 store under oxygen; N_2 store under nitrogen; L+ store under light; L- store protected from light; T increase temperature (data from Wells (1988)).

Decomposition seen	Mechanism	Action to mitigate
All	Hydrolysis	Control pH or reduce water content
O_2 but not N_2	Oxidation	Package under N_2 or add chelator/antioxidant
O_2L+ but not O_2L-	Photolysis	Protect from light
O_2T and N_2T only	Thermal	Cool or refrigerate

decomposition as well as some of the indicative actions that might be taken to minimise degradation.

Photostability testing is perhaps the hardest test to accomplish, because of the difficulty in standardising the spectral distribution and intensity of light to which the sample is exposed. ICH Guideline Q1B (1996) defines the light source, and light boxes are available, but ensuring uniform exposure of the sample is complicated by the fact that light sources change as they age; powdered samples will also be exposed to light more on one side than the other.

A rapid method for determining the likelihood of oxidation if a nitrogen bag is not available is to prepare two samples in solution, one with oxygen being bubbled through it and the other with a metal chelator added. A marked improvement in stability in the presence of the chelator indicates oxidation as the mechanism of degradation.

9.5.2   Stress testing in the solid-state

Stability should be tested as a function of temperature, using the stress conditions specified in ICH Guideline Q1A(R2)(2003). These require temperatures above those used for accelerated stability testing (40 °C), although room-temperature storage (or cold storage at 4 °C) is useful, if only to justify that any extrapolation from high-temperature data is justified. A typical storage and testing scheme is shown in Table 9.5. Thermal analysis (particularly DSC) or XRPD are useful for analysis post stress, since they will detect changes in physical form while an HPLC assay can be used to detect loss of the drug substance, giving an insight into chemical stability. Stressing the sample by exposure to light may well result in a colour change, in which case photostability is implied before any quantitative analysis is attempted. Water uptake as a function of RH is also useful, as is determination of stability at a number of RH values, since water is ubiquitous. As an aside, variation in the stability data generated during stress testing may often be traced back to variability in the water content of the sample, so where as much as practicably possible samples should be desiccated prior to use or stored under defined RH (saturated salt solutions will maintain defined RH's in a closed

Table 9.5 A typical storage and testing protocol for preformulation stress testing (data at the lower temperatures are useful to check the validity of data extrapolation from high temperatures).

	Storage period						
	Hours		Days				
Temperature (°C)	6	24	3	7	14	21	28
4						+	+
25					+	+	+
37			+	+	+	+	+
50		+	+	+	+	+	+
60		+	+	+	+	+	
70	+	+	+	+			

environment; see Table 9.6. The salt solution may be placed in the base of a desiccator, providing a closed environment of defined RH).

It is worth compacting the drug substance, since this will bring the material into greater physical contact, although this may also reduce exposure to light and humidity. Stressing the sample with a mechanical force (such as by ball milling) is also a useful indicator of physical stability, since processing in this way may produce new surfaces or surface defects. It will also indicate the likelihood that the material will turn partially amorphous during processing.

9.5.3 Drug–excipient compatibility testing

Obviously a formulated medicine is very different in terms of physical form and chemical composition from a pure drug substance. It can be surprising how many other excipients are present in what would appear to be a simple medicine. For instance, a 200 mg Nurofen (ibuprofen) tablet contains 12

Table 9.6 The relative humidities maintained by various saturated salt solutions at 15, 20 and 25 °C (data from Greenspan (1977)).

	RH Maintained		
Salt	15 °C	20 °C	25 °C
Lithium chloride	11.3	11.3	11.3
Magnesium chloride	33.3	33.1	32.8
Potassium carbonate	43.1	43.2	43.2
Sodium bromide	60.7	59.1	57.6
Sodium chloride	75.6	75.7	75.3
Potassium chloride	85.9	85.1	84.3
Potassium sulphate	97.9	97.6	97.3

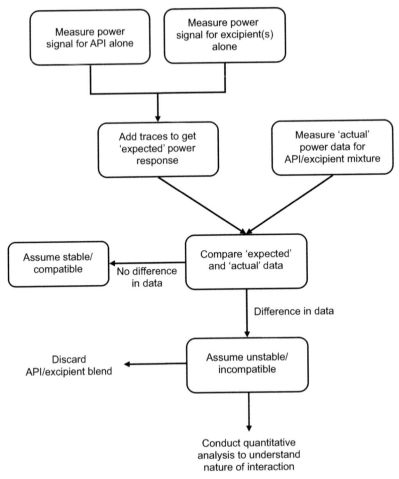

Figure 9.14 Work-flow diagram for performing a drug substance–excipient compatibility screen with DSC.

excipients (sucrose, sodium citrate, talc, croscarmellose sodium, stearic acid, titanium dioxide, silicon dioxide, acacia, carmellose sodium, sodium lauryl sulphate, macrogel and black ink).

It is imperative for the preformulation scientist to appreciate that while such additives may be pharmacologically inert (often denoted as being generally regarded as safe, GRAS), they are often far from inert physicochemically. Thus, the successful formulation of a stable and effective dosage form depends on the careful selection and testing of drug–excipient combinations. At the preformulation stage, the final dosage form may not have been decided, but checking the compatibility of the drug substance with a range of potential excipients will give an early indication of which excipients to exclude from the drug product.

There are a number of ways of performing excipient compatibility screening, although the basic concept is always to blend the drug substance and excipient together, place under stress conditions and analyse. Typically, two components are dry-blended (preferably ensuring that the particle sizes of the two materials are the same) in a 1:1 mass ratio. Once initial screening has been completed, more appropriate mass ratios can be used. As in the case of solid-state stress testing, the mixture can be compressed prior to storage, which will bring the components into greater physical contact, or analysed as a powder blend. The analysis is performed on the individual components and the mixture and any unexpected events or products, in the case of the physical blend, provides an indication that some interaction may have occurred (Figure 9.14).

Again, DSC and HPLC are particularly useful for analyses post stress. It can be argued that only isothermal testing under storage conditions will satisfactorily predict (with any degree of confidence) the real situation. However, stress testing is a practical necessity, especially during preformulation, and 3 weeks storage at 50 °C (which is kinetically equivalent to 12 weeks at ambient temperatures) is a generally accepted compromise. Other useful screens for solid-state stability are XRPD and FTIR.

Great caution should be taken in interpretation of the data, since compatibility screens are designed to be the worst-case scenario and so may not reflect the case of the formulated drug product. A good example is magnesium stearate, which often reacts with drug substances when present in high mass ratios. These results do not, however, preclude its widespread use as a tablet lubricant at levels of 0.25–0.5% w/w.

Study question 9.6   Discuss any drawbacks of using DSC for drug substance–excipient compatibility testing.

Summary box 9.5

- Stability assessment during preformulation is focused on stress conditions.

- Initial degradation in strong acid and base will identify hydrolysis.

- Matrix design involving exposure of solution to heat, light and oxygen will identify degradation pathways.

- Solid-state stability assessed with DSC.

- Excipient compatibility testing will inform drug product design.

9.6 Summary

Understanding the stability profile of a drug substance, both in isolation and formulated as a drug product, is critical in ensuring safety, efficacy and quality. ICH Guidelines provide stability testing protocols for regulatory submission, covering storage, intermediate, accelerated and stress conditions. At the preformulation stage, the aim of stability testing is rapidly to identify any potential degradation pathways and to get an indicative mechanism – determination of long-term degradation kinetics and stability profiles comes later in the development process. Thus the focus is on stress conditions, using a minimum amount of drug substance. Susceptibility to hydrolysis should be determined as a function of pH, since hydrolysis reactions are frequently acid or base catalysed. If hydrolysis is suspected, the control of water content during processing, formulation and storage will be critical. Oxidative degradation is easily screened for, either by storing the sample under an inert atmosphere (typically nitrogen) or adding a metal chelator or antioxidant to the solution. Photosensitivity testing is required for all drug substances and is covered by ICH Guideline Q1B (1996). During preformulation characterisation, exposure of the sample to strong light intensity will be sufficient to highlight photodegradation if any is occurring. Determination of the causative wavelength can be determined at a later stage and used to inform any strategy to improve stability.

References

Connors, K.A., Amidon, G.L. and Stella, V.J. (1986) *Chemical Stability of Pharmaceuticals*, 2nd edn, John Wiley & Sons, Ltd. ISBN 0-4718-8795-X.

Day, J.N.E. and Ingold, C.K. (1941) Mechanism and kinetics of carboxylic ester hydrolysis and carboxyl esterification. *Trans. Faraday Soc.*, **37**, 686–705.

Fryklöf, L.-E. (1954) Autoxidation of etheric oils used in pharmacy. *Farm. Rev.*, **53**, 317–335.

Greenspan, L. (1977) Humidity fixed of binary saturated aqueous solutions. *J. Res. Natl Bur. Stand.*, **81A** (1977) 89-96

Grimm, W. (1986) Storage conditions for stability testing (Part 2). *Drugs Made in Germany*, **29**, 39–47.

Grimm, W. (1998) Extension of the International Conference on Harmonisation Tripartite Guidelines for stability testing of new drug substances and products to countries of Climatic Zones III and IV. *Drug Dev. Ind. Pharm.*, **24**, 313–325.

ICH Harmonised Tripartite Guideline Q1A(R2) (2003) Stability testing of new drug substances and products.

ICH Harmonised Tripartite Guideline Q1B (1996) Stability testing: photostability testing of new drug substances and products.

Ingold, C.K. (1953) *Structure and Mechanism in Organic Chemistry*, Cornell University Press. ISBN 0-8014-0499-1.

ISO 10977 (1993) Photography. Processed photographic colour films and paper prints – methods for measuring image stability.

Kommanaboyina, B. and Rhodes, C.T. (1999) Trends in stability testing, with emphasis on stability during distribution and storage. *Drug. Dev. Ind. Pharm.*, **25**, 857–868.

Kondrat'eva, A.P., Bruns, B.P. and Libinson, G.S. (1971) Stability of *d*-cycloserine in aqueous solutions at high concentrations. *Pharm. Chem. J.*, **5**, 356–360.

Loucas, S.P., Mehl, B., Litwak, R.S. and Jurado, R.A. (1981) Stability of procaine hydrochloride in a buffered cardioplegia formulation. *Am. J. Hosp. Pharm.*, **38**, 1924–1928.

Wells, J.I. (1988) *Pharmaceutical Preformulation. The Physicochemical Properties of Drug Substances*, John Wiley & Sons, Ltd, Chichester. ISBN 0-470-21114-8.

Answers to study questions

9.1 Unfortunately, if the drug has poor solubility and is being developed as a salt, the pH of maximum stability may well also be the pH of minimum solubility. Also the pH of maximum stability for an ophthalmic drug may not be at the optimum pH for transcorneal absorption, or may be well removed from neutrality, causing local irritation.

9.2 Options to reduce the extent of oxidative degradation in the drug product would be to design a packaging system to minimise exposure to oxygen or to add an antioxidant (such as ascorbic acid or sodium metabisulphite) or metal chelator (such as EDTA or ethylene diamine). Discolouration can be masked with a coloured or opaque film coat.

9.3 Starting with Equation (9.6),

$$[A]_t = [C]_0 - kt$$

The shelf life is the time taken for 10% degradation to occur or for $[A]_0$ to fall to 90% of its initial value:

$$[A]_{t0.9} = 0.9[A]_0$$

Substitution yields

$$0.9[A]_0 = [A]_0 - kt_{0.9}$$

and hence

$$t_{0.9} = [A]_0/10k$$

A similar analysis can be performed for the half-life.

9.4 Following a similar logic to the zero-order case,

$$\ln[A]_t = \ln[A]_0 - kt$$

$$\ln 0.5[A]_0 = \ln[A]_0 - kt_{0.5}$$

$$kt_{0.5} = \ln[A]_0 - \ln 0.5[A]_0$$

$$kt_{0.5} = \ln\{[A]_0/0.5[A]_0\}$$

Assuming $[A]_0 = 1$, $kt_{0.5} = \ln 2$ or $t_{0.5} = 0.693/k$, and for the shelf-life, $t_{0.9} = 0.1054/k$.

9.5 The change from first-order reaction kinetics in solution to zero-order kinetics in the solid-state is because in the latter state the sparsity of water present usually means that the water film around the solid particles is saturated with water-soluble ingredient(s). Thus, as degradation reactions continue in this water film, more ingredient dissolves into the layer of water, thus keeping it saturated. Therefore the concentration of the decomposing drug in solution remains constant and hence zero-order kinetics are observed.

9.6 The basic principle of using DSC to determine compatibility or incompatibility is that a DSC trace of a nonreacting powder mixture will be the sum of the component ingredients. Thus comparison of the trace produced by the mixture with the traces of the individual components will show the presence or absence of any interactions. However, as usual, things are not that simple, and there are occasions when the simple act of measuring a mixture will result in some slight changes in shape of the DSC trace that are not as a result of any detrimental interaction. Interactions can manifest themselves in changes in the position of melting point peaks, peak shape and peak area, and the possible appearance of a transition not shown by any of the individual components. For example, there could be the emergence of new peaks and the gross broadening or elongation of others. Also the presence of second-order transitions (which produce changes in the baseline) will indicate a chemical interaction of some type between the drug and the excipient or the production of a eutectic (a solid-solution melt). Any small indication of such evidence indicates that the drug and excipient are chemically incompatible and that the particular excipient–drug combination should be avoided totally in any future product. In some cases where observed thermal interactions are small yet an interaction is suspected, any potential incompatibility can be confirmed (or otherwise) by TLC and/or HPLC.

10
Particle Properties

10.1 Introduction

Since manufacturing processes frequently involve the movement, blending and manipulation of solids, understanding the properties of powders is an essential precursor to successful formulation. The microscopic structures of crystalline and amorphous materials have already been discussed, and it is important to note that the macroscopic behaviour of particles is not always predictable from crystallographic data. This is because other factors, such as particle size, shape, surface roughness and surface chemistry predominate. The best way to characterise particles is by visual observation, typically with a light microscope,[1] although electron and atomic force microscopies are useful when the particle size is small. This chapter will focus on particle characterisation. Particle–particle interactions and powder flow are discussed in Chapter 11.

10.2 Microscopy

As usual with analytical instrumentation, the operating principles of each type of microscope vary as do the nature and type of samples that can be visualised. Most pharmaceutical powders are in the 1–3000 μm range, although formulation of nanomedicines is becoming an increasingly popular strategy. The efficacy of many pulmonary devices is also critically dependent upon particle size, with powders needing to be in the 2–5 μm range for effective pulmonary delivery. In preformulation, light microscopy and electron microscopy are the most common instruments used to characterise samples.

---

[1]Takeru Higuchi reportedly said that a light microscope was his favourite research tool.

Essentials of Pharmaceutical Preformulation, First Edition. Simon Gaisford and Mark Saunders.
© 2013 John Wiley & Sons, Ltd. Published 2013 by John Wiley & Sons, Ltd.

10.2.1　Light microscopy

The term microscopy derives from *micros* (small) and *skopeo* (look at). The inventor of the microscope is unclear, but an early design by Antonie van Leeuwenhoek (1632–1723) was used to discover protozoa and bacteria.[2] Modern microscopes do not differ fundamentally from the design of van Leeuwenhoek, but use more advanced optics to magnify the image. All light microscopes contain:

- A light source

- A condenser lens

- An objective lens

- An eyepiece (or ocular)

In a typical configuration there are two light sources, one under the sample (for transmission mode) and one above the sample (for reflectance mode); selection of the light source depends on the thickness (or transparency) of the sample. The choice of lenses affects the magnification of the microscope, which in turn affects the *resolving power* (defined as the minimum distance between two resolvable points – the unassisted eye has a resolving power of 0.2 mm). The resolving power is related to the numerical aperture (NA) of the objective lens and the wavelength of light (λ) used to illuminate the sample:

$$\text{Resolving power} = \frac{\lambda}{\text{NA}} \qquad (10.1)$$

The numerical aperture is dependent upon both the lens used and the refractive index (n) of the medium through which the light is transmitted:

$$\text{NA} = n \sin \mu \qquad (10.2)$$

where μ is the angle of the cone of light produced by the objective lens. It should be apparent that the resolving power of a light microscope is thus limited by the wavelength of visible light (which is why electron microscopies have much greater resolving powers, see below). It should also be clear that the resolving power can be increased by increasing the numerical aperture of the objective lens, which can be achieved by increasing the refractive

---

[2]There are two microbiology journals that carry his name: *The Antonie van Leeuwenhoek International Journal of General and Molecular Biology* and *The Antonie van Leeuwenhoek Journal of Microbiology*.

index of the medium between the sample and the lens. Air has a refractive index of 1 and the maximum NA attainable is 0.95. If the medium is changed to a mineral oil, the maximum NA attainable is increased to 1.6. The maximum magnification attainable with a light microscope is of the order of $\times 1000$.

One further consideration is depth of field (the distance in front of and behind the focal point that appears to be in focus). This will reduce with increasing magnification, which means that unless samples are quite thin, it can be difficult to get the whole sample into focus.

10.2.2 Hot-stage microscopy

A very important derivative of light microscopy is hot-stage microscopy (HSM). Here, a standard light microscope is fitted with a temperature-controlled stage, which means that the sample can be visualised as it is heated and cooled at a defined rate. Recall from Chapter 7 that DSC in particular is a very widely used tool in preformulation but that the data can be difficult to interpret in isolation. Visualisation of the changes occurring in a sample as a function of temperature is a powerful aid to interpretation of events seen in DSC thermal curves. It was noted earlier that operating HSM with cross-polarising filters was a particularly useful analytical method. This is because polymorphs are birefringent and so appear coloured. How is this so?

The wave model of light assumes vibration at right angles to the direction of propagation, with each vibration direction being equally probable. A plane-polarising filter allows only the light vibrating in one direction to pass through it. A cross-polarised light microscope contains two plane-polarising filters, one between the light source and the sample (the polariser) and the other between the sample and the ocular lens (the analyser). The filters are oriented at 90° to each other initially, although the operator can usually rotate the polariser. Ordinarily, then, this arrangement prevents transmission of any light from the source to the ocular lens. Crystalline materials are usually birefringent, which means the plane-polarised light is transformed into two wave components (the *ordinary* and *extraordinary* wavefronts) polarised in mutually perpendicular planes. The velocities of the wavefronts are different, and so out of phase, but they are recombined with constructive and deconstructive interference by the analyser, producing a coloured image. Different unit cells will interact with plane-polarised light to varying extents and so will produce patterns of different colours. Amorphous materials, conversely, are not birefringent and so do not orient plane-polarised light in any particular way, and appear black. This is a quick and useful check to determine whether a sample is amorphous.

While TGA can be used to quantify mass changes with temperature, these events can often also be seen with HSM. The sample is surrounded with a liquid in which it will not dissolve (usually a mineral or silicone oil) – any process that results in a gaseous degradant will be seen as a release of bubbles.

Study question 10.1   How might cross-polarised HSM be useful in interpreting DSC data?

10.2.3   Electron microscopy

From Equation (10.1) it can be seen that the resolving power of a microscope is dependent upon the wavelength of light used to illuminate the sample. Using a beam of electrons as the 'light source' greatly increases the resolving power of the microscope because the wavelength at which electrons travel in a vacuum is much shorter (ca. $\times 10\,000$ shorter). Electron beams cannot be focused with optical lenses, so electromagnets are used instead. Neither can they (nor should they) be visualised by eye, so fluorescent screens are used. Other than this change, the operating principles of electron microscopes are very similar to the light microscopes discussed above. As for light microscopes, electron microscopes can be operated in transmission (TEM) or reflectance (scanning or SEM) modes.

In TEM the electrons must pass through the sample and so the sample must be extremely thin. Typical applications include imaging nanoparticles, viruses and liposomes.

In SEM the electron beam is focused onto a 1–5 nm spot on the sample surface. The majority of electrons will pass through the sample but some interact with and are then emitted from atoms in the sample and are deflected back to the detector. These are known as backscattered or secondary electrons. The quality of the image is dependent upon the number of secondary electrons. To increase the number, samples are commonly covered in a heavy metal (such as gold) prior to imaging. Newer instruments do not operate under such a high vacuum and use a cascade of electrons generated from water molecules in the atmosphere to generate an image; these are called environmental SEM (ESEM) instruments. In either case, the electron beam is moved in an x–y direction across the sample surface in a raster pattern to construct the image.

10.2.4   Atomic force microscopy

Atomic force microscopy (AFM) works on completely different principles from the optical or electron microscopies described above. The principal part of an AFM is a small cantilever with a sharp tip. The cantilever is typically constructed from silicon or silicon nitride and the size of the tip is a few (1–10) nm. The size of the tip governs the image resolution. The tip is held in close proximity to the surface of the sample. On the top side of the cantilever is a small mirror, on to which a laser is aimed, the laser beam being deflected on to a detector.

In scanning mode, the cantilever is moved in an x–y direction across the surface of the sample in a raster pattern. As the topography (or surface roughness) of the sample changes, so does the height of the cantilever – movement in this z dimension changes the position of the laser beam on the detector, and so a topographical image of the sample surface can be constructed.

In tapping mode the cantilever is oscillated at or near to its resonance frequency by a small piezoelectric actuator (the amplitude of oscillation is typically 100–200 nm). As the tip nears the surface of the sample, forces of interaction (such as van der Waal's forces, dipole–dipole interactions and electrostatic forces) change the amplitude of oscillation. An electronic servo uses the piezoelectric actuator to control the height of the cantilever above the sample and acts to maintain constant oscillation amplitude. A surface image can then be constructed by moving the cantilever in an x–y direction over the sample.

Summary box 10.1

- Light and electron microscopies operate on similar principles, but electron microscopies have higher resolving powers because of the shorter wavelength of electron beams.

- Thin samples can be imaged in transmission mode.

- Thick samples are imaged in reflectance mode. Light microscopes suffer from a narrow depth of field.

- Atomic force microscopy uses the deflection of a cantilever to build up a topographical image of a sample surface.

10.3 Particle shape

10.3.1 Habit

It was noted in Chapter 7 that crystalline materials are defined as having a unit cell that repeats in three dimensions to produce the macroscopic crystal and that the gross exterior appearance of the crystal is called its habit. Six basic habits are listed in the USP (acicular (needle-shaped), blade, columnar, equant (equal-sided or isometric, including cubic and spherical), plate and tabular), although other terms are frequently encountered (such as hexagonal, octahedral, prismatic and pyramidal). Since the names are just descriptive terms for the macroscopic shape of the crystal and do not necessarily relate to the geometry of the unit cell, it can be difficult to identify polymorphic or pseudopolymorphic forms by image analysis alone. However, knowledge of habit is important, because particle shape can affect processability (influencing factors such as compressibility, flowability, mixing and filtration). Similarly, particle size is often critical to drug product performance and habit can change as the crystal grows in solution.

Since the molecules in a crystal lattice are all arranged in the same orientation, it follows that different faces of a crystal will have different properties (since different functional groups will be exposed at each face), such as surface energy, polarity or chemical reactivity. Further, because crystals grow by condensation of molecules (from the melt or from solution) it may well be the case that condensation is more favoured on certain faces and less favoured on others, with the outcome that the crystal may grow preferentially in certain directions (anisotrophy).

The effect of preferential growth on certain faces of the crystal is manifest in changes in habit. Figure 10.1 shows what will happen to a hexagonal seed crystal if growth occurs on different faces. When growth occurs on faces a and d the final habit is a diamond, while when growth occurs on faces d and f the

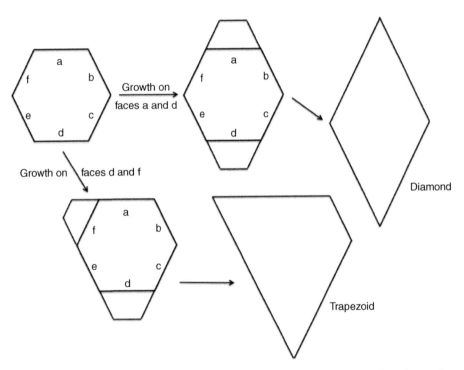

Figure 10.1   Effect of preferential growth on different faces of a hexagonal seed crystal.

final habit is a trapezoid. This is why simply observing the crystal habit does not allow assignment of a unit cell structure. Note here that those faces upon which crystal growth is preferred eventually disappear while at the same time those faces upon which growth is less favoured become larger. The net result is that the *largest face is always the slowest growing.*

If crystal growth were to occur in a polar solvent (like water), formation of polar surfaces will be favoured (and so condensation will occur on nonpolar faces, since these will then become smaller) and vice versa. This means that habit may be controlled by solvent selection. Rasenack and Müller (2002) prepared two ibuprofen and five acetaminophen crystal habits by selection of crystallisation solvents. In the case of ibuprofen thin plates showed the best compaction, while for acetaminophen prismatic crystals were preferable. The growth rate may also affect habit (and will also vary depending upon solvent selection), with fast growth rates often resulting in acicular crystals (noted by Banga *et al.*, 2007, for crystallisation of celecoxib).

Other processing factors that may affect the crystal habit produced include:

- Supersaturation. Excessive supersaturation tends to lead to prismatic or tabular crystals.

- Cooling rate. The rate of cooling affects the rate of change of supersaturation. An example is naphthalene, wherein rapid cooling produces thin plates, while slow cooling results in prisms.

- Co-solvents. The addition of co-solvents, other solutes or ions can change habit by poisoning growth from a particular face. For instance, sodium chloride grows as cubes from water but in the presence of trace urea gives an octahedral habit.

Similarly, any factor that affects the face upon which molecules condense can potentially affect habit. Surfactants, for instance, may affect crystal habit (and so they are often termed *crystal habit modifiers*). The reason is because surfactants will adsorb to certain faces of the crystal preventing condensation. For example, Kaul, Nguyen and Venkataram (1992) showed that addition of PVP, PEG 4000 or gelatin to a supersaturated solution of an orally active metal chelator resulted in a change in habit from orthorhombic to spherical.

Finally, crystal growth may occur in *planes*, which are not necessarily equivalent to faces. Planes are named using the system of *Miller indices* – three of the planes available in a cubic unit cell are shown in Figure 10.2. Considering the plane shown on the left, the plane cuts both the *a*- and *c* axes at one unit length but never cuts the *b* axis (an intercept of ∞). Miller indices are written as the reciprocal of the three intercepts; in this case $a = 1, b = 0$

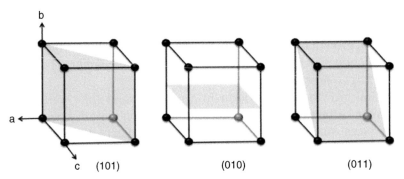

b

a

c   (101)   (010)   (011)

Figure 10.2   Three of the possible planes (and Miller indices) in a cubic unit cell.

and $c = 1$ and so the plane is named (101). A similar analysis shows the plane on the right to be (011). The plane in the centre never intersects the a and c axes but bisects the b axis (so has intercepts $a = \infty$, $b = 0.5$, $c = \infty$). Taking reciprocals gives $a = 0$, $b = 2$ and $c = 0$. Now another rule is applied: to divide all the indices by the largest common factor (2 in this case), which gives $a = 0$, $b = 1$ and $c = 0$ and so plane (010). Two other rules may be applied depending on the plane being named. One is to remove any fractional reciprocal indices by multiplying with a suitable factor and the other is to indicate negative reciprocals with a bar above the number. In this fashion, all possible planes can be named.

Preferential growth on planes can have the effect of dramatically modifying the habit. Figure 10.3 shows the growth and change in habit of an orthorhombic crystal. During preformulation the synthetic route to a drug candidate may not have been optimised, so a range of polymorphs or habits may be produced. These need to be identified and characterised for the

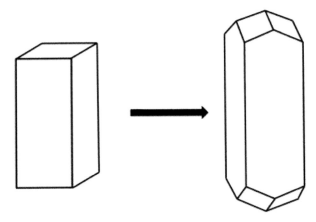

Figure 10.3   Effect of preferential growth on crystal planes on the habit of an orthorhombic crystal.

best combination of stability, bioavailability and processability. Clearly, polymorph selection is the priority, since this will affect stability and bioavailability, followed by habit selection, which will primarily affect processability.

Summary box 10.2

- Crystal shape is called habit. Six basic habits are defined.

- The surface chemistry of different faces of a crystal will be dependent upon the functional groups exposed.

- Growth on certain faces can be encouraged by choice of solvent and can result in a change in habit.

- The largest face is always the slowest growing.

- Growth can also occur on planes (named with Miller indices).

- Surfactants can act as crystal habit modifiers by binding to selected faces of a crystal.

10.3.2 Particle sizing

Having some idea of the particle size distribution (a sample is extremely unlikely to have a monodisperse particle size distribution) of a sample can be important both for processability (powder flow, mixing, etc.) and for drug product performance (inhalers). There are many techniques for particle sizing but all have limitations and it is important to understand that the size distributions produced are only as good as the sample property data (density, refractive index, etc.) used in any algorithm. Typical sizing techniques include:

- Coulter counter, or electrical sensing zone (ESZ). This is a method based on the changes in conductivity of an electrolyte as particles in suspension are drawn through an aperture. The change in conductivity is thus a function of the volume of electrolyte displaced. ESZ (Coulter) diameters are calculated by assuming that the particles are spherical.

- Photon correlation spectroscopy (PCS). This involves placing a dispersion of the powder in a fluid (either air or a nondissolving liquid) in a parallel beam of laser light. The light is diffracted by the particles and detected with a diode array; the smaller the particle, the greater the angle of diffraction, and vice versa. Particle sizes are again calculated on the basis that all particles are spherical.

- Sieve analysis. A series of sieves of increasing pore size are arranged in a vertical stack (largest at the top) and the powder placed at the top. The sieves are then shaken and the fraction of sample retained in each sieve is calculated from the weighed mass of material retained in each sieve.

All these techniques require considerable sample amounts and are thus not really available to the preformulation scientist. The best option is thus visual inspection of a sample with microscopy. The particle sizes recorded in this way are accurate and the particle shape can be visualised, but it is difficult to construct a distribution and only a small fraction of the sample is considered. Also, defining the shape of a particular sample can be difficult, as real samples are not generally composed of crystals of equal geometry. The same problem applies to powders, which may have been produced by mechanical particle size reduction or blending. Some common particle shape descriptors are given in Figure 10.4.

Once the shape has been determined, determination of particle size is complicated if the shape is not uniform. Various methods of defining particle size have been proposed (Brittain, 2001). One concept is to compare habits

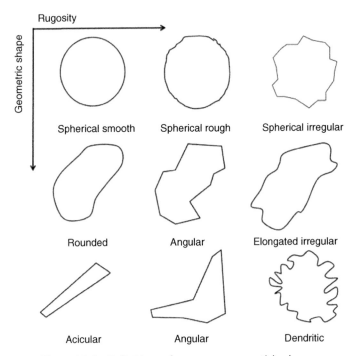

Figure 10.4 Definitions of some common particle shapes.

Table 10.1 Surface areas for three geometric shapes of equivalent volume.

Shape	Dimensions (μm)	Volume (μm^3)	Surface area (μm^2)
Sphere	30	113 040	11 304
Cube	48.4 (all sides)	113 380	14 055
Needle	$10 \times 10 \times 1130$	113 000	45 400

in terms of their aspect ratio (AR, the ratio of the length to the width). An AR less than 5 usually means flowability problems are avoided, while an AR approaching 1 (i.e. equant) means good processability.

Habit may affect dissolution rate because different geometric shapes will have different surface area to volume ratios. Table 10.1 shows the surface areas for three geometric shapes with (roughly) equivalent volumes. It is apparent that an acicular needle has a much greater surface area than a cube (3.2-fold greater) or a sphere (4-fold greater). This means that acicular needles will have a faster dissolution rate than an equivalent crystal form of cubic or spheroidal habit. As ever, there is thus a compromise to be reached between good processability (low AR) and dissolution (high AR).

If the particles are imaged with a light microscope, it is possible to use an ocular that has a series of rings drawn on it, of calibrated size, to which particles may be referenced. Again, it may be difficult to determine the exact shape of a particle, because the light microscope image is generated in two dimensions but the particle exists in three dimensions or because the particles are not spherical – some statistical approaches to match an irregular particle to a defined circle are shown in Figure 10.5. The use of digital cameras to

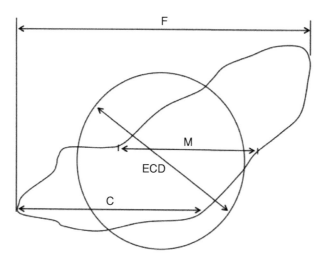

Figure 10.5 Some statistical measures of particle size for an irregular particle measured against a standard circle. Key: C, maximum horizontal chord (or intercept); ECD, equivalent circle diameter; F, Feret's diameter; M, Martin's diameter.

capture images also means software algorithms can be used for particle size characterisation.

More commonly, SEM is used for particle size estimation, because it does not suffer from depth of field issues and it can image a wide range of particle sizes. Typically, five samples from the whole are taken and imaged. In each image there should be around 200 particles. An SEM image is produced with a calibrated ruler and hence visual analysis can determine particle sizes. Alternatively, software can be used to determine the fraction of an image containing particles by analysis of pixel colours.

Summary box 10.3

- Knowing particle shape is the first step in defining particle size.

- It is difficult to define particle size for irregular shapes.

- Images taken with a light microscope can be referenced against a standard circle.

- Alternatively, definitions based on aspect ratios can be used.

- Several instruments are available for measuring particle size, but direct visualisation of the sample is the safest option.

10.3.3   Particle size distributions

Particle size data are usually presented in terms of the number (or frequency) of particles within a given size range (a frequency distribution histogram, given in Figure 10.6). A limitation of plotting data as a frequency distribution histogram is the difficulty in defining a single mean diameter that is representative of the whole powder. This problem is particularly acute when the distribution is not symmetrical (in other words, the mean particle size is not the median).

One solution is to convert the data into a cumulative distribution (either cumulative percentage undersize or oversize). In this case the median size of the powder will be the size at which 50% of the particles are larger (oversize) and 50% are smaller (undersize). Since it is very unlikely that this point will coincide exactly to a measured experimental point, graphical interpolation is usually required (Figure 10.7).

Notice that in Figure 10.6 there are two curves, one for cumulative percentage oversize by weight and the other by number and that the distributions (and so the mean sizes) are different. This is important, because pharmaceutical powders are compacted or dispensed by weight (equivalent to dose) yet

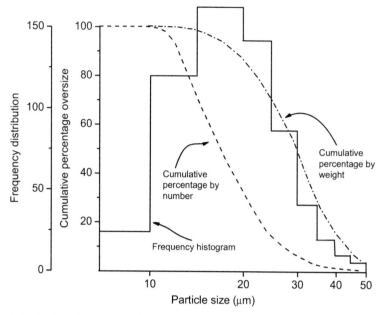

Figure 10.6 Particle size data as a frequency distribution histogram and cumulative percentage by number and weight.

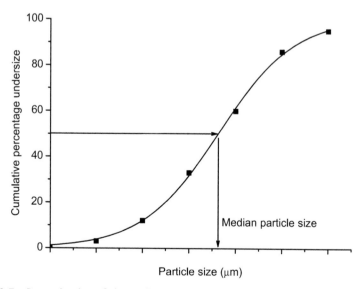

Figure 10.7 Determination of the median diameter of a powdered material from cumulative percentages.

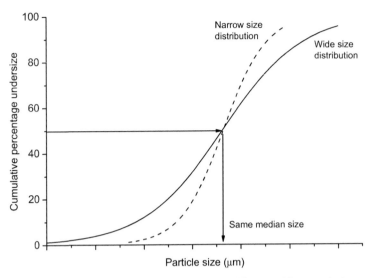

Figure 10.8 Narrow and wide particle size distributions illustrated by cumulative percentage undersize plots.

their physiological performance is more than likely going to depend in some way on size distribution. For instance, a sample with a small median particle size will have a greater surface area and so appreciably faster dissolution than a sample with a large median particle size. Equally, the presence of a few large crystals (high mass, low surface area) in an otherwise fine powder would not seriously influence the number distribution but would likely have a serious effect on the overall surface area.

While cumulative plots allow accurate quantification of mean size, they do not quantify the extent of scatter in the particle sizes. Knowledge of whether the particle size distribution is narrow or wide is important because, as noted above, the coarse fraction may exert considerable influence over the macroscopic performance of the powder. Scatter can be assessed visually in a semi-quantitative manner from cumulative plots, since a narrow distribution gives a much steeper curve than a wide distribution (Figure 10.8).

With all powders, the majority of particles will be of a size close to the mean; as the distance from the mean is increased the number of particles will reduce. Statistically, an ideal (or normal) distribution of sizes would result in a bell-shaped (or *Gaussian*) curve. Experimentally, however, most powders tend to have a relatively higher proportion of smaller particles and fewer large particles, resulting in a non-Gaussian (or skewed) distribution (Figure 10.9).

A skewed distribution is often rendered Gaussian if particle size is plotted on a log scale (a *log normal* distribution). It can be shown that for a Gaussian distribution the area under the curve between the mean and ± 1 standard deviation accounts for 68% of the particles (i.e. between 16 and 84%). The

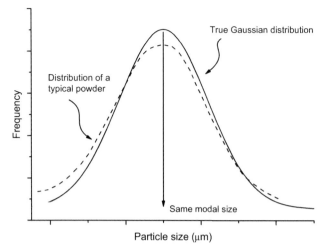

Figure 10.9   Gaussian (or normal) distribution relative to a typical powder distribution.

data at these points, together with that at 50%, can be used to calculate what is called the geometric standard deviation (σ_g):

$$\sigma_g = \frac{\text{Size at 16\% oversize}}{\text{Size at 50\%}} \tag{10.3}$$

$$\sigma_g = \frac{\text{Size at 16\% undersize}}{\text{Size at 50\%}} \tag{10.4}$$

The application of these equations is shown in Figure 10.10. A distribution that in pharmaceutical terms could be described as narrow would have a σ_g value of about 1.5 while wide distributions would have values of 2 or greater. Particles that were exactly monodisperse would have a geometric standard deviation of 1.

Summary box 10.4

- Particle shape determined by visual inspection of a sample is not necessarily representative of distribution of sizes with the whole sample.

- Particle size distributions can be presented as a frequency histogram, but determination of the mean value is difficult.

- Cumulative under- or oversize plots ameliorate this issue, as a 50% value is known.

- Ideal distribution is symmetrical about the mean but in practice pharmaceutical powders tend to be skewed in favour of smaller particles.

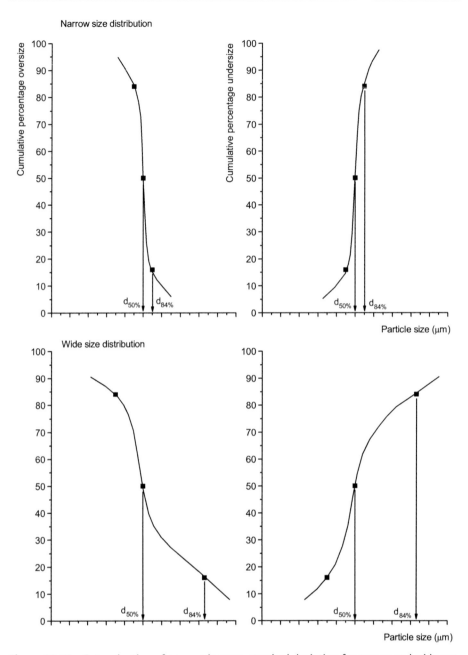

Figure 10.10   Determination of geometric mean standard deviation for narrow and wide particle size distributions.

10.4 Summary

While molecular packing exerts great control over the stability and dissolution rate of a drug substance, particle shape plays a critical role in how the material will process. A given polymorph may grow into various habits, depending upon solvent polarity and the addition of crystal habit modifiers.

Habit is best determined by visual inspection of the sample, typically with light microscopy, although electron microscopies can be used for smaller particles. Atomic force microscopy is more suited to determination of surface properties, such as topography and surface energy. Image analysis can also be used to determine particle size distribution, although this method assumes that the fraction of sample visualised (small) is representative of the whole. Alternatively, particle sizing apparatus can produce distribution data.

References

Banga, S., Chawla, G., Varandani, D. and Mehta, B.R. (2007) Modification of the crystal habit of celecoxib for improved processability. *J. Pharm. Pharmacol.*, **59**, 29–39.

Brittain, H.G. (2001) Particle size distribution, Part 1, representations of particle shape, size and distribution. *Pharm. Technol.*, **25**, 38–45.

Kaul, D., Nguyen, N.T. and Venkataram, S. (1992) Crystal habit modifications and altered tableting characteristics. *Int. J. Pharm.*, **88**, 345–350.

Rasenack, N. and Müller, B.W. (2002) Crystal habit and tableting behavior. *Int. J. Pharm.*, **244**, 45–57.

Answer to study question

10.1 The problem with DSC data is that many events can give rise to similar data (noted in Chapter 7). Hence, measuring the same sample under the same heating (or cooling) rates with complementary techniques is a powerful aid to interpretation. The most obvious events visible with HSM will be phase transitions such as melting, polymorphic change and crystallisation since these occur with a change in physical form (either to a liquid or with change in particle shape). If conversion between polymorphs upon heating is suspected from the DSC data then heating the sample on the hot stage of the microscope should result in changing colour patterns being observed at the temperature of any exothermic peaks. Changes in habit can be differentiated from changes in polymorphic form because the particle shape will change while the colour pattern will not.

11
Powder Properties

11.1 Introduction

The bulk properties of a powder, such as flow and compressibility, are easily overlooked during preformulation, when poor properties are of no real consequence. However, on the manufacturing scale powder properties are critical, since any powder or powder blend must flow into tablet presses or capsule machine hoppers smoothly and, most importantly, uniformly. Powder blend uniformity must remain constant after any manufacturing step since any change will result in the batch of drug product failing content uniformity testing. Although manufacturing processes are clearly outside the scope of preformulation, attention paid to characterising powder properties at this stage can significantly reduce the chances of problems in scale-up later in development.

Powder flow is governed by many factors, including temperature, water content, particle shape and particle–particle interactions. The complex interplay between these factors means that absolute prediction of powder flow from measurement of particle properties is virtually impossible and flow characteristics of the bulk powder must be made directly. Such measurements are unlikely to be made during preformulation because of the limited amount of material available, but flow can be estimated by measurement of bulk density and the angle of repose and understood by measurement of the forces of cohesion and adhesion.

11.2 Powder flow and consolidation

Powders flow when the forces acting on them are enough to overcome the forces of adhesion or cohesion between particles. In other words, there is a

Essentials of Pharmaceutical Preformulation, First Edition. Simon Gaisford and Mark Saunders.
© 2013 John Wiley & Sons, Ltd. Published 2013 by John Wiley & Sons, Ltd.

position of equilibrium:

$$\sum F_{\text{driving}} = \sum F_{\text{drag}} \qquad (11.1)$$

Factors that contribute to driving forces include gravitation, particle mass, the angle of inclination of the powder and any applied load, while factors that contribute to drag forces include forces of adhesion or cohesion, electrostatic forces, water bridges and mechanical interlocking. Good flow would mean the powder would move under gravity, without the need for any additional excipients. In quantitative terms, flow means that the powder mass deforms plastically in response to an applied load (or stress).

In general, particle size greatly influences flow properties and is the easiest variable to change during processing. Fine powders, with high surface area-to-mass ratios, tend to be affected more by the force of cohesion. Coarse particles tend to be affected more by gravitational forces. As a general rule, particles larger than 250 μm are usually free flowing, but cohesive forces start to predominate when the particle size reduces to 100 μm or less. When the particle size is as small as 10 μm, powders are usually very cohesive.

The density of a powder is defined by the ratio of its mass to volume; however, unlike pure solids, powders consist of solid particles with air (or void) spaces between them. Thus, while the individual particles have a true density, the measured density of the powder will be smaller, as a result of the void spaces. The term *bulk density* is used to describe the density of the powder and its value must be smaller than the true density of the particles. It follows that bulk density may vary. For instance, if the particle size distribution of a sample decreases, the ratio of void space to particles will also decrease, increasing the bulk density. In this case, bulk density must be defined for a powder with a certain particle size distribution. Further, if a powder is poured into a container, it is likely that initially the particles are at their least ordered (i.e. the ratio of void space to particles will be at its highest and the bulk density at its lowest). This is defined as the 'poured' or 'fluff' bulk density. Tapping the sample will result in reduction of the bulk volume as the particles reorder themselves (termed *consolidation*). The ratio of void space to particles will be at its lowest and the bulk density at its highest. This is called the 'tapped' or 'final' bulk density.

Determination of fluff and tapped bulk density indicates the degree to which a powder will consolidate under gravity. If an additional load is applied vertically to the powder, further consolidation will occur. For a consolidated powder to flow, the gravitational force must be sufficient to overcome the forces of consolidation and cause plastic deformation.

A number of indices have been developed to relate the degree of consolidation to powder flow properties. The two most commonly used are those

of Carr (1965) and Hausner (1967). In addition, the relationships between
stresses and powder flow can be visualised with a Mohr diagram.

Summary box 11.1

- Whether a powder flows or not depends on the balance between driving
  forces and drag forces.

- Drag forces are influenced by particle size and shape.

- Particles greater than 250 μm tend to be free flowing, while those below
  100 μm tend to be cohesive.

- Particles have a true density, but the density of the bulk powder will be
  smaller, because of void spaces.

- A freshly poured powder will have a lower bulk (fluff) density than a
  consolidated powder (tapped density).

11.2.1   Carr's index

Carr (1965) developed a simple test using a jolting volumeter. This apparatus consists of a measuring cylinder located vertically in a frame with a cam
below it; as the cam rotates, the measuring cylinder is jolted up and down,
simulating tapping. The speed of rotation of the cam and the number of taps
are controlled by an electronic interface and motor. The number and force
of each tap are thus precisely controlled. A typical experimental apparatus is
shown in Figure 11.1.

The powder is sieved initially, rather than poured, into the cylinder so
that it is at its maximum porosity and has not been partially compressed by
the filling process. Then all that is necessary is to measure the volume of a
given weight of a bulk powder before tapping to obtain the fluff density and
again after vertically jolting the cylinder to give the tapped density.

When large quantities of material are available, a 100 mL measuring
cylinder is employed. Clearly, during early development phases much less
drug candidate may be available, so the equipment may be modified with a
smaller measuring cylinder. One option would be to mount the small cylinder
through a rubber bung and position it in the neck of the larger cylinder.

The initial volume is recorded and called V_0. Then the contents are jolted
and the volume of the powder recorded after 2, 4, 6, 8, 10, 12, 15, 20, 30 and
finally 50 taps (V_f). Weighing the cylinder full and again empty will give the

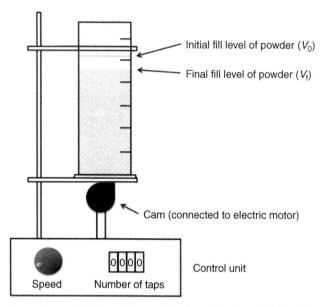

Initial fill level of powder (V_0)

Final fill level of powder (V_f)

Cam (connected to electric motor)

Control unit

Speed   Number of taps

Figure 11.1 An automated jolting volumeter for determination of fluff and tapped density.

weight of powder (W). From this, the fluff density (g mL^{-1}) can be calculated from W divided by V_0 and the tapped density from W divided by V_f.

Carr quantified the degree of 'compressibility' as

$$\text{Compressibility (\%)} = \frac{\text{Tapped density} - \text{Fluff density}}{\text{Tapped density}} \times 100 \quad (11.2)$$

The relationship between Carr's compressibility index and the probable flow properties of a powder mass is shown in Table 11.1.

Table 11.1 Relationships between Carr's compressibility index, Hausner ratio and powder flow.

Carr's compressibility index (%)	Hausner ratio	Description of flow
<10	1.00–1.11	Excellent
11–15	1.12–1.18	Good
16–20	1.19–1.25	Fair
21–25	1.26–1.34	Passable
26–31	1.35–1.45	Poor
32–39	1.46–1.59	Very poor
>40	>1.60	Very, very poor

11.2.2 Hausner ratio

A similar index was defined by Hausner (1967). The Hausner ratio is defined as

$$\text{Hausner ratio} = \frac{\text{Tapped density}}{\text{Fluff density}} \qquad (11.3)$$

The relationship between the Hausner ratio and powder flow is given in Table 11.1. Because both Carr's and Hausner's ratios are one-point determinations, calculated from the initial and final volumes of the powder, they do not indicate the ease and speed with which consolidation occurs. Some materials might have quite a high index (suggestive of poor powder flow) but in fact compress rapidly. This is very useful for uniform die filling on tablet machines when the powder may flow into the die close to its minimum density and then quickly compresses to maximum density prior to consolidation. An empirical linear relationship exists between the percentage bulk density and the log of the number of taps. Nonlinearity occurs up to 2 taps and then after 30 taps when the bed compresses more slowly. Between these points the relationship should be linear, the slope being a measure of the speed of compression. The value is useful for assessing powders or blends of powders of similar indices, the beneficial effects of glidants and/or the design of capsule formulations.

11.2.3 Angle of repose

If a mass of powder is poured on to a flat surface it will form a cone with a characteristic angle to the horizontal surface. Assuming the powder is free flowing, the angle will be small and a shallow cone will be formed. Conversely, where the powder exhibits poor flow the angle will be high. The angle, known as the angle of repose (θ), can therefore be used to predict powder flow even when only a small amount of material is available.

Experimentally, a mass of powder is placed in a funnel (the base of which is blocked with a rubber bung) over a metal disc. The bung is removed and the powder falls on to the disc, forming a heap. The diameter of the base of the heap is known, and a constant, because the size of the metal disc is known.

If the height (h) and diameter (D) of the cone are determined, θ can be calculated (Figure 11.2):

$$\tan \theta = \frac{h}{D/2} \qquad (11.4)$$

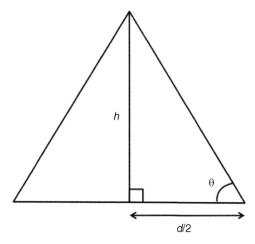

Figure 11.2 Determination of the angle of repose from a powder cone.

Measurements of θ are notoriously variable, but it is still possible to relate θ to flow, as shown in Table 11.2. It is most important to ensure that the top of the cone is not rounded by powder falling and impacting on the top of the cone.

Study question 11.1 Can you see the problem of inaccurate measurement of θ generated by rounding the top of the cone? Can you suggest a test method that would not deform the cone?

Study question 11.2 Can you suggest how you might adapt existing or standard angle of repose determinations when only a small weight of powder is available?

Table 11.2 Relationship between angle of repose and powder flow.

Angle of repose (θ)	Description of flow
<25	Excellent
25–30	Very Good
31–35	Good
36–40	Fair
41–45	Passable but flow aid might be needed
46–55	Poor – agitation or vibration needed
>56	Very poor

In general, powder behaviour can be split into three categories:

- Low Carr's index, low angle of repose. Flow is good so no future problems are likely.

- Mid-range values for Carr's index and angle of repose. Flow can probably be improved with additives.

- High values of Carr's index and angle of repose. Flow is extremely poor and cannot be improved sufficiently with the use of additives.

Study question 11.3   Can you suggest which additives may be useful in improving the flow of powders with mid-range values for Carr's index? What might be done for powders with high Carr's index values?

Study question 11.4   Explain how the angle of repose and powder flow are related in terms of interparticulate cohesion.

Figure 11.3 summarises the interrelationship between Carr's compressibility index, the angle of repose and flowability of a powder.

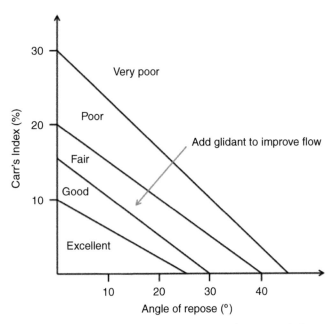

Figure 11.3   Relationship between Carr's index and angle of repose and the consequence for powder flowability.

Summary box 11.2

- Carr's index and the Hausner ratio are two methods used to define the compressibility of a powder and either can be measured with a tapping volumeter.

- Angle of repose is derived from a powder heap – lower angles imply better flow.

- The correlation of Carr's index with angle of repose gives a good indication of bulk powder flow properties.

11.2.4 Mohr diagrams

Mohr diagrams are a visual way of representing the relationships between the stresses on a powder and its flow properties and as such provide useful quantitative parameters for characterisation. Understanding the construction of Mohr diagrams starts with defining the forces that act on a powder at rest. For any given plane, the total force exerted is the sum of forces acting perpendicular to the plane (normal stress, σ) or parallel to the plane (shear stress, τ). Figure 11.4 shows a mass of powder held within a cylinder (for the purpose of this discussion, there is assumed to be no friction between the cylinder walls and the sample. Also, it is assumed that there are no shear stresses acting on the top or bottom surfaces of the powder). A normal (i.e. vertical) force

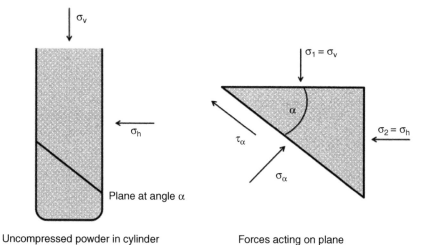

Uncompressed powder in cylinder   Forces acting on plane

Figure 11.4 Forces acting on an uncompressed powder contained in a cylinder (left) and the forces acting on a plane through the powder (right).

(σ_v) acts on the powder (and so $\sigma_v > 0$). Were a powder to act like a Newtonian fluid, the forces acting on it in all directions would be equal. In practice, powders do not act like Newtonian fluids and so there will be stresses acting on the sample in every possible plane, which will comprise both normal and shear components. The force acting on the horizontal plane is defined as σ_h. The ratio σ_h/σ_v is termed the stress ratio (K) and is itself a value with which to characterise powders (K is typically between 0.3 and 0.6 for pharmaceutical powders).

The forces acting on any other possible plane (at an angle α from the horizontal) may be calculated by considering a triangular cross-section of the powder bed (Figure 11.4). From this the following relationships can be calculated (Schulze, 2008):

$$\sigma_\alpha = \frac{\sigma_v + \sigma_h}{2} + \frac{\sigma_v - \sigma_h}{2} \cos(2\alpha) \qquad (11.5)$$

$$\tau_\sigma = \frac{\sigma_v - \sigma_h}{2} \sin(2\alpha) \qquad (11.6)$$

A plot of σ_α versus τ_α for all possible values of α will result in a circle (Figure 11.5). The circle (which will have a radius of $(\sigma_v - \sigma_h)/2$ and a centre at $(\sigma_v + \sigma_h)/2$) represents the normal and shear stresses acting on the powder at all possible planes and is known as a Mohr diagram. The circle will always

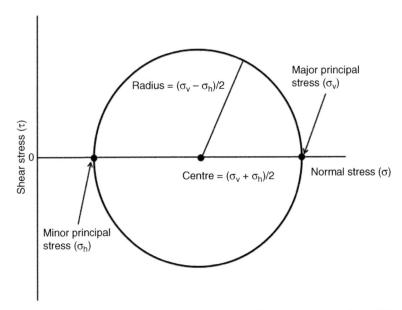

Figure 11.5   A Mohr diagram representing the normal and shear forces for all possible planes through a powder.

have two points of intersection with the x axis, corresponding to two values of normal stress. The larger value is termed the major principal stress (σ_1) and the smaller value the minor principal stress (σ_2). Making the assumption that no shear stresses are acting on the top and bottom of the sample, and that the walls are frictionless, σ_v must be the major principal stress and σ_h the minor principal stress.

11.2.4.1   Mohr diagrams and consolidation   The powder sample repre- sented in Figure 11.4 was freshly poured and so no consolidation has yet occurred. Imagine that the powder is consolidated, by application of a ver- tical stress (a consolidation stress, or major principal load). The powder will compress; the smaller the final volume, the greater the compressibility of the powder. This action represents the compression of a powder mass into a com- pact or tablet. If the consolidation stress is removed, the powder will remain compacted. If the compact, now removed from the cylinder, is loaded with an increasing vertical stress, at some point it will fracture (called incipient flow). The stress required to cause incipient flow is equal to the compressive strength of the compact. This point is also called the yield point of the com- pact; below the yield point the compact does not flow and above the yield point the compact flows under plastic deformation.

A Mohr diagram can be constructed for consolidation. During consolida- tion a normal stress is acting on the surface of the powder and a correspond- ing (but smaller) stress acts on a horizontal plane (as determined by the ratio K). There are no shear stresses and so the Mohr diagram will be a circle (sym- metrically located on the x axis, since $\tau = 0$) with two points of intersection with the x axis – semi-circle C on Figure 11.6.

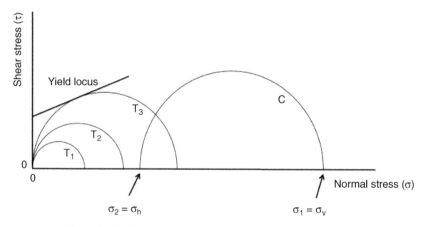

Figure 11.6   Mohr circles (only the positive shear stress values are shown) for an uncompressed powder (C) and three samples after a vertical force has been applied. T_1 and T_2 are below the yield value while the yield value has been reached at T_3.

Now consider the situation during compression testing. The powder (now a compact) is removed from the cylinder and a vertical stress (less than the yield value) is applied. The major principal stress is the vertical stress and the minor principal stress is the horizontal stress (which is now zero, as the sides of the compact are no longer constrained by the cylinder). The Mohr diagram will look like semi-circle T_1 on Figure 11.6. Since the minor principal stress is zero, one point of intersection is the origin. Similar Mohr diagrams can be drawn for increasing vertical stresses up to the yield value (semi-circles T_2 and T_3). Again, the minor principal stress will be at the origin. Semi-circle T_3 represents the case where the yield value has been reached and incipient flow has occurred. Because the compact fractured, the yield value must have been reached along at least one of the planes through the sample, and so no larger Mohr circle is possible (because once the sample has fractured flow can occur and so no greater force can be applied). A line connecting all the Mohr circles represents the yield locus of the sample. In principle, the point of intersection of the yield locus with the y axis is equivalent to the force of cohesion (or adhesion if the powder is a blend) of the powder, because at this point the normal stress applied is zero. Were the force of cohesion or adhesion zero, the yield locus would pass through the origin.

11.2.4.2   Determination of the yield locus   The discussion above assumed uniaxial compression of a powder bed to determine the Mohr diagrams and, consequently, the yield locus. Although fine in principle, in practice it is extremely difficult to construct an experiment such that frictionless walls are obtained. Also, the yield locus is derived from the gradient of a line tangential to the Mohr circle where incipient flow occurred, but determination of the exact tangent is rather subjective. For these reasons, shear testers are usually used to determine the yield locus.

In essence, all shear testers follow the same principle. A powder mass is consolidated between two plates by application of a vertical load (σ). The top plate is then moved horizontally at a constant rate (v), resulting in the sample experiencing a horizontal shear stress (τ). With increasing shear stress the resulting force (F) acting on the sample increases until incipient flow is seen.

There are some important experimental considerations. Firstly, the sample should be consolidated under a preload (σ_{pre}) sufficient that the shear stress it experiences is constant (Figure 11.7). At this point, neither compact strength nor bulk density increase any further with an increase in load and the sample is considered critically consolidated with respect to normal stress. This also means that other samples from the same powder, consolidated with the same σ_{pre} should behave in the same manner. Before the top plate is moved horizontally, an experimental shear load (σ_{sh}) is placed on the sample, where σ_{sh} must always be less than σ_{pre} (such that if σ_{sh} had been used

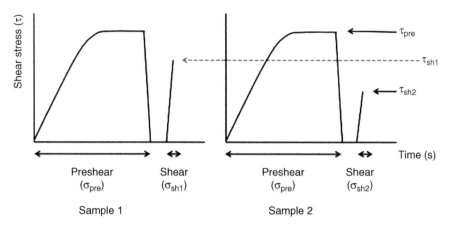

Figure 11.7   Plots of shear stress as a function of time for two samples undergoing shear testing. The samples are consolidated under the same preload stress, but sheared under different normal loads.

as the preload, the sample would not have been critically consolidated). A shear force is then applied by moving the top plate horizontally until the sample flows. The point of maximum shear stress immediately prior to incipient flow (τ_{sh}) corresponds to the yield point. The result is a value of σ_{sh} and a corresponding value of τ_{sh}. The experiment is repeated (on a new compact) with a range of values of σ_{sh}, enabling construction of a plot of σ_{sh} versus τ_{sh} (Figure 11.8). The yield locus is the slope of the line.

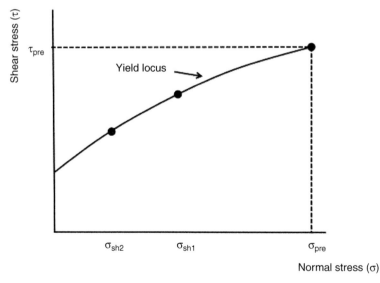

Figure 11.8   Shear stress versus normal stress plot for consolidated samples showing construction of the yield locus.

> **Summary box 11.3**
>
> - Mohr diagrams relate the stresses applied to a powder with its yield locus (or breaking point).
>
> - When a compact breaks under stress the break point is termed incipient flow and occurs at the yield locus.
>
> - The yield locus is measured with a shear tester.
>
> - Compacts are prepared under the same preload force but sheared under different (and smaller) normal loads.

11.3 Compaction properties

Unfortunately, the compression and cohesion properties of most drug substances are generally very poor and so excipients are required in order to manufacture acceptable compacts. With low-dose, high-potency drug substances, the compaction characteristics of the drug are less important because the majority of the compact will be an excipient or excipient blend, chosen to have good compressibility. When the dose is significant, however, say greater than 50 mg, the compressibility of the drug substance cannot be ignored and will greatly influence the overall properties of the compact. Again, while processing is beyond the scope of preformulation, an initial assessment of compressibility will aid selection of the drug substance selected for development as well as give an early indication of the likely need for any excipients.

The ideal substance for compression would have a combination of plastic deformation and brittleness. Plastic deformation means that the substance will flow under a compressive load and will remain deformed once the load is removed. Brittleness indicates that a substance will fracture, creating new surfaces that may encourage adhesion or cohesion. In all cases, the water content of the powder is critical, since water can act either as a lubricant or plasticiser, as is consideration of whether the sample is wholly or partially amorphous.

Sebhatu, Elamin and Ahlneck (1994) show the effect on tablet strength for a series of compacts of lactose (nominally 15% amorphous, prepared by spray-drying). The lactose was stored at 57% RH for various periods of time prior to compression. At up to 4 hours of storage, compact strength was observed to increase with storage time. At longer time periods, however, compact strength was reduced to a minimum (Figure 11.9). Similarly, larger water contents were seen to correlate with greater strength.

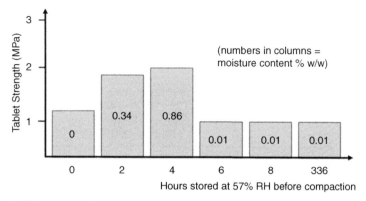

Figure 11.9 Tablet strength as a function of storage time (at 57% RH) for spray-dried lactose (redrawn from Sebhatu *et al.* (1994) with kind permission from Springer Science+Business Media B.V.).

Study question 11.5   Can you think of a reason for the trend in tablet strength with storage time?

Assuming that excipients are needed in the powder blend then a useful practical rule is that if a high dose drug behaves plastically, the excipients should fragment. Similarly, if the drug is brittle or elastic, the excipients should deform plastically. Wells (1988) suggested a scheme to assess whether or not a material has elastic, plastic, fragmentary and punch-filming properties when even only a small amount of material is available, by compacting only three tablets:

- Three aliquots of drug substance (500 mg) are weighed. To each is added magnesium stearate (5 mg) as a lubricant.

- Two of the aliquots are blended with magnesium stearate for 5 min (A and B) while the third is tumble-mixed for 30 min (C).

- Samples A and C are loaded into a 13 mm IR punch and die-set and compressed quickly to 1 ton for 1 s. The compacts are stored in a sealed container overnight.

- Sample B is compressed in the same manner but the load is applied for 30 s.

- The crushing force for each compact is measured.

Interpretation of the results is given in Table 11.3.

Study question 11.6   Can you see and explain the logic of this scheme?

Table 11.3   Interpretation of the compression data suggested by Wells (1988).

	Type of material	
	Plastic	Fragmenting
	Comparison of crushing force	
Compare strengths of compacts A and B	A < B	A = B
Compare A and C	C < A	A = C
Overall	C < A < B	A = B = C

11.3.1 Compaction simulators

An alternative, rapid, method to determine the compaction properties of a powder is to use a compaction simulator. Such an apparatus compacts single tablets under conditions that equate to those that would be experienced in a commercial tablet machine.

Typically a compaction simulator consists of two vertically aligned high-speed hydraulic rams. A punch is attached to each ram and, together with a die, form the same arrangement used in tablet machines. The hydraulic rams can move at speeds that approximate those of even the fastest commercial rotary tablet machines. Their most attractive feature, however, is that changes in the velocity of the punches can be controlled, by computer, in such a way that they can exactly match the compaction cycle of any commercial tablet machine. Unlike commercial tablet machines, compaction simulators can produce single tablets and so are very useful during preformulation.

Summary box 11.4

- Most drug substances do not exhibit satisfactory compression properties.

- Ideally a substance for compression would deform plastically and have a degree of brittleness.

- If the dose is low then excipients can be selected with good compression properties.

- If the dose is high then the properties of the excipients must be matched to those of the drug substance.

- If the drug substance deforms plastically the excipient should be brittle and vice versa.

Testing in this manner is particularly important when the material has time-dependent properties. Such materials may compact satisfactorily on a hydraulic press or on a slow, single-punch tablet machine (of the type that may be found in development laboratories), but when attempts are made to run these tablets on a high-speed production machine, the duration of the compaction (dwell time) is too short for sufficient deformation to occur in order to produce a strong tablet.

11.4 Summary

Powder flow properties are not important during preformulation character-isation and may not significantly influence the decision on which drug sub-stances are selected for development, but there is no doubt that poor powder flow will be a considerable challenge that must be overcome during scale-up if a successful drug product is to be manufactured. With only limited amounts of sample available, tests on powder flow are limited, but good insight can be gained by measuring compressibility and angle of repose. Each measurement is quick, requires little powder and does not destroy or change the sample, yet the data give a rapid indication of the flow properties and likely need for excipients.

References

Carr, R.L. (1965) Evaluating flow properties of solids. *Chem. Engng*, **72**, 163–167.

Hausner, H.H. (1967) Friction conditions in a mass of metal powder. *Int. J. Powder Metall.*, **3**, 7–13.

Jones, T.M. and Pilpel, N. (1966) Some angular properties of magnesia and their relevance to material handling. *J. Pharm. Pharmacol.*, **18**, 182T.

Schulze, D. (2008) *Powders and Bulk Solids. Behaviour, Characterisation, Storage and Flow*, Springer, Berlin. ISBN 978-3-540-73767-4.

Sebhatu, T., Elamin, A.A. and Ahlneck, C. (1994) Effect of moisture sorption on tableting characteristics of spray dried (15% amorphous) lactose. *Pharm. Res.*, **11**, 1233–1238.

Wells, J.I. (1988) *Pharmaceutical Preformulation. The Physicochemical Properties of Drug Substances*, John Wiley & Sons, Ltd, Chichester. ISBN 0-470-21114-8.

Answers to study questions

11.1 The impact of the falling powder on to the top of the cone rounds off the top. Normally the angle of the cone is not measured directly, but is calculated after measuring the height of the resulting cone and diame-ter of its base using the equation $\tan \theta = h/(D/2)$, where θ is the angle of repose (i.e. the angle between the horizontal base and sloping side of the cone. Thus, if the top of the cone has been reduced in height, h will be incorrect (too short) and the calculated value of θ will be too low.

This will give the impression that the powder has better flow properties than is really the case.

The problem of cone deformation can be avoided by using the method of Jones and Pilpel (1966). This involves putting the powder under test into a cylinder that is closed at the bottom apart from a small circular hole, which is temporarily sealed with a rubber bung. The cylinder also contains a horizontal circular platform midway up the cylinder. When the bung at the bottom is removed the powder flows out of the hole, leaving a perfect, undistorted cone on the circular platform. Since the diameter of the cone (D) is fixed (the diameter of the platform) all that is required by the test is to measure the height of the top of the cone above the platform. The same equation can now be used.

11.2 This question provides another example of how the preformulation scientist must adapt standard techniques to cope with the small amounts of material available. A miniature version of the Jones and Pilpel apparatus as described in the answer to Question 11.1 above could be constructed but an alternative is to measure the 'angle of spatula'. This requires an excess of powder to be poured on to the end of a spatula blade. Simple observation will reveal whether the cone is steep (poor flow) or shallow (good flow).

11.3 Glidants are added to powders to improve their flow properties. Glidants are not to be confused with lubricants, which are added to reduce die wall friction and punch sticking during tablet compaction. Some materials are efficient at both jobs while others may be good glidants yet poor lubricants. Magnesium stearate is very frequently used, in concentrations between 0.25 and 1.0% w/w. Talc is used much less frequently nowadays but is a moderately good lubricant and glidant when used in concentrations between 1 and 2% w/w. Other pharmaceutically relevant glidants are colloidal silica (0.1–0.5% w/w) and starch (2–10% w/w).

Powders with extremely poor flow properties cannot be rescued by the addition of lubricants alone and steps must be taken to reduce significantly interparticulate attraction. Since this is essentially a problem of cohesion, reducing the specific surface area of the powder can lead to improved flow. Powder agglomeration, and thus size enlargement, by granulation or roller compaction is the best and most successful approach. Increasing the size of the particles will reduce their specific surface area and so reduce the total cohesive force. Coincidentally, granulation will improve the shape of the particles since numerous irregular particles are bundled together into a roughly spherical (isodiametric) mass.

11.4 A powder particle at rest on a static horizontal surface will be totally stable. If the surface is tilted slightly, there will be no movement of that particle since it is held in place by attractive forces of adhesion between the powder particle and the surface. If the angle of at which the surface is tilted is increased, the particle will eventually slide or roll down the surface. Similarly, imagine a powder particle sitting on a horizontal layer of similar particles. It will again be stable until an angle of tilt of the bed is reached at which the interparticulate forces can no longer hold that particle in place and it will roll down the surface of the bed. The angle of the powder bed so formed (the angle of repose) will be dependent on the magnitude of interparticulate friction. The greater the value of interparticulate friction the greater will be the angle of the powder bed before the particle begins to roll. Thus powders with high interparticulate friction will have a high angle of repose, and vice versa.

11.5 What is happening is that the amorphous fraction of the sample (15% by mass) is absorbing water during storage at 57% RH. This is shown by the increasing water content up to 4 h storage. The water plasticises the amorphous material, making it deform more plastically under compression, and so the compact is compressed to a smaller volume with a greater strength. After 4 h storage, however, the amorphous material crystallises, so the remaining tablets are made by compressing crystalline lactose. This undergoes less plastic deformation and so produces a weaker tablet. There is not a straight correlation with water content, because the initial sample (also essentially dry) compresses to a higher strength than the crystalline lactose. This is because the initial sample is 15% by mass amorphous material, and amorphous material undergoes better plastic deformation than crystalline material. If there was not good control of storage of the lactose post spray-drying then there would be significant batch-to-batch variability in the strengths of the tablets produced.

11.6 When materials are ductile, they deform by changing shape (plastic flow). Since there is no fracture, no new surfaces are generated during compression and thus a more intimate mix of magnesium stearate (C) leads to poorer bonding. Since the deformation of these materials is time-dependent, increasing the dwell time at consolidation (B) will increase bonding and compact strength. If a material is predominantly fragmenting, neither addition of lubricant, increased mixing time (C) or increased dwell time (B) should affect the tablet strength.

Some materials, such as paracetamol, are predominantly elastic and little permanent change occurs during compaction as the crystals return to their original dimensions on load removal. This results in weak compacts. If this bonding is also very weak the tablet can 'self-destruct' and the top will detach (capping) or horizontal cracks will occur throughout the compact (lamination). An elastic material made according to scheme A will cap or laminate, one made according to B probably maintain integrity but be very weak and by scheme C will cap or laminate. Elastic drugs require a plastic tableting excipient or wet massing to induce plasticity.

Finally, the surface of the top and lower punches should be examined for the adhesion or powder (punch filming, sticking). The punches can be dipped into a suitable extraction solvent and the drug level determined. This may be higher for A and B since magnesium stearate is an effective anti-adherent and 30 minutes of mixing (C) should produce a monolayer of magnesium stearate over the drug particles and thus suppress adhesion more effectively.

This is yet another excellent example of how to achieve the maximum amount of extremely important data from a minimum number of experiments on expensive and scarce material.

Index

Essentials of Pharmaceutical Preformulation, First Edition. Simon Gaisford and Mark Saunders.
© 2013 John Wiley & Sons, Ltd. Published 2013 by John Wiley & Sons, Ltd.